W0017519

Developing an Army Strategy for Building Partner Capacity for Stability Operations

Jefferson P. Marquis, Jennifer D. P. Moroney, Justin Beck,
Derek Eaton, Scott Hiromoto, David R. Howell, Janet Lewis,
Charlotte Lynch, Michael J. Neumann, Cathryn Quantic Thurston

Prepared for the United States Army
Approved for public release; distribution unlimited

RAND ARROYO CENTER

The research described in this report was sponsored by the United States Army under Contract No. W74V8H-06-C-0001.

Library of Congress Cataloging-in-Publication Data

Developing an Army strategy for building partner capacity for stability operations / Jefferson P. Marquis ... [et al.].
    p. cm.
  Includes bibliographical references.
  ISBN 978-0-8330-4954-4 (pbk. : alk. paper))
      1. United States. Army—Stability operations. 2. Integrated operations (Military science) I. Marquis, Jefferson P.

    U167.5.S68D48 2010
    355.4—dc22

                                                                      2010017395

The RAND Corporation is a nonprofit research organization providing objective analysis and effective solutions that address the challenges facing the public and private sectors around the world. RAND's publications do not necessarily reflect the opinions of its research clients and sponsors. **RAND®** is a registered trademark.

Cover: U.S. Army photo by Kenneth Fidler

Published 2010 by the RAND Corporation
1776 Main Street, P.O. Box 2138, Santa Monica, CA 90407-2138
1200 South Hayes Street, Arlington, VA 22202-5050
4570 Fifth Avenue, Suite 600, Pittsburgh, PA 15213-2665
RAND URL: http://www.rand.org/
To order RAND documents or to obtain additional information, contact
Distribution Services: Telephone: (310) 451-7002;
Fax: (310) 451-6915; Email: order@rand.org

Preface

This report documents the results of a project entitled "Building Partner Capacity for Stability, Security, Transition and Reconstruction Operations." The purpose of the project was to assist the U.S. Army and other U.S. government agencies in their efforts to develop a well-defined and integrated BPC for stability operations strategy based on empirical analysis.

The research reported here was sponsored by the Office of the Deputy Chief of Staff, G-3, Headquarters, Department of the Army. The research was conducted in RAND Arroyo Center's Strategy, Doctrine, and Resources Program. RAND Arroyo Center, part of the RAND Corporation, is a federally funded research and development center sponsored by the U.S. Army. The report includes information that was available to the authors as of mid-2007. Consequently, the study does not include more recent Office of Secretary of Defense (OSD) and Joint Staff guidance on building partner capacity for stability operations and how that affects the Army.

This report should be of interest to those concerned with the impact of security cooperation programs designed to build the capacity of partner nations to conduct stability operations in a coalition or indigenous context.

Jefferson Marquis and Jennifer Moroney are the lead authors. To contact them for comments or further information: Jefferson Marquis (telephone 310-393-0411, extension 6123, Jefferson_Marquis@rand.org); Jennifer Moroney (telephone 703-413-1100, extension 5940, Jennifer_Moroney@rand.org).

The Project Unique Identification Code (PUIC) for the project that produced this document is DAMOC07199.

For more information on RAND Arroyo Center, contact the Director of Operations (telephone 310-393-0411, extension 6419; fax 310-451-6952; email Marcy_Agmon@rand.org), or visit Arroyo's web site at http://www.rand.org/ard/.

Contents

Figures

Tables

Summary

Secretary of Defense Robert M. Gates has defined the war on terror as "a prolonged, worldwide irregular campaign—a struggle between the forces of violent extremism and those of moderation." According to Gates, in order to effectively carry out such a campaign, the military must learn two hard lessons from the wars it has conducted in Afghanistan and Iraq since the fall of 2001.

First, "over the long term, the United States cannot kill or capture its way to victory."[1] In other words, "soft power"—including diplomacy, strategic communications, foreign assistance, civic action, and economic reconstruction—is at least as important as, if not more than, "hard power" in creating the conditions for the eventual defeat of violent extremism throughout the world.[2] To better meet this challenge, Gates has called for an increase in the capacity of civilian national security agencies—in particular, the U.S. Department of State (DOS) and the U.S. Agency for International Development (USAID)—so they can take the lead in exercising soft power in unstable parts of the globe. In addition, the Defense Secretary has recognized that the Department of Defense (DoD) must continue to play a major role in stability operations—maintaining security, providing humanitarian aid, beginning

---

[1]   Robert M. Gates, "A Balanced Strategy: Reprogramming the Pentagon for a New Age," *Foreign Affairs*, January/February 2009.

[2]   "U.S. defense chief urges greater use of 'soft power,'" Agence France-Presse, November 26, 2007.

reconstruction, bolstering local governments and public services, espe-cially "in the midst of or in the aftermath of conflict."[3]

Gates's second major strategic lesson from recent U.S. inter-ventions in the Middle East is the desirability of taking an "indirect approach" to prosecuting the war on terror. In his view, because the United States is unlikely to mount another major invasion and occupa-tion in the foreseeable future, it should follow a sustainable counterter-rorism strategy that does not rely on the massive application of U.S. combat power. Ideally, the United States should work "by, with and through" its allies and partners and, when necessary, bolster the capac-ity of their governments and security forces to effectively contribute to the war on terror.

Study Purpose and Approach

The U.S. government is facing the dual challenge of building its own interagency capacity for conducting stability operations while simulta-neously helping to build partner capacity (BPC) for stability operations across a wide range of nations. The purpose of this study is to assist the U.S. Army, DoD, and other U.S. government agencies in developing a well-defined, well-integrated BPC for stability operations strategy and to create a nexus between the concepts of BPC and stability opera-tions. To accomplish this goal, a RAND Arroyo Center study team conducted an exploratory analysis of key strategic elements within the context of BPC and stability operations guidance as well as ongoing security cooperation programs, using a variety of analytical techniques.

Concepts

As currently conceived, BPC is a multi-agency, multinational initiative that draws on the elements of security cooperation to achieve U.S. stra-tegic objectives. To help achieve these objectives, both U.S. allies and partners can act as force multipliers and as a hedge against future secu-

---

[3] Gates.

rity requirements. With greater global demands for U.S. forces and an expanding list of adversaries, conditions, and crises that could threaten U.S. national interests, allies and partners increase and diversify the capabilities needed to counter a range of threats on unfamiliar geographical and cultural terrain.[4]

"Stability operations" is an evolving and variously named concept. Historically, the U.S. military tended to relegate operations that do not involve full-scale combat to several overlapping but not identical categories: small wars; low-intensity conflicts; military operations other than war; small-scale contingencies; peace operations; stability and support operations; stability, security, transition, and reconstruction (SSTR) operations; or simply stability operations. Despite their differences, all of these concepts refer to military operations in civilian environments. According to DoD Directive 3000.05, military support for SSTR operations consists of DoD activities "that support U.S. Government plans for stabilization, security, reconstruction and transition operations, which lead to sustainable peace while advancing U.S. interests."[5] In the interest of brevity, we use the Army's "stability operations" term throughout this report.

Findings and Recommendations

For this study, we conducted an exploratory analysis of five strategic elements necessary to align U.S. government security cooperation efforts with the goal of BPC for stability operations in a largely peacetime environment. Figure S.1 lays out the organization of the strategic elements as well as the corresponding chapter in which each element is examined in depth.

---

[4] White House, *The National Security Strategy*, September 2002, p. 29.

[5] Headquarters, Department of the Army (HQDA), *Stability Operations*, FM 3-07, 2008; and Department of Defense Directive 3000.05, *Military Support for Stability, Security, Transition, and Reconstruction Operations*, November 28, 2005. Although valid for the period of our study, this directive was reissued as Department of Defense Instruction 3000.05, *Stability Operations*, September 16, 2009. The new instruction updates policy and assigns responsibilities for the identification and development of DoD capabilities to support stability operations.

Figure S.1
BPC for Stability Operations: Strategic Elements with Related Analyses

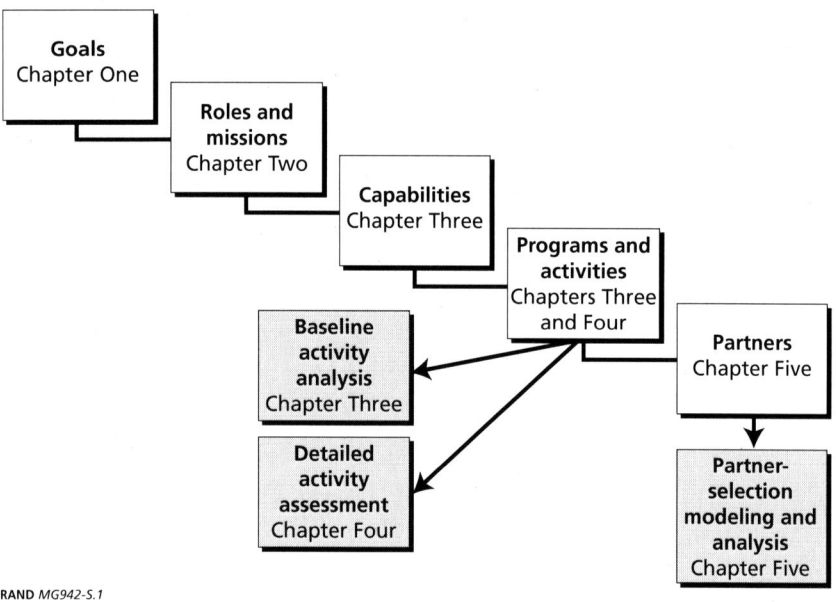

RAND *MG942-S.1*

To integrate the five elements essential to BPC for stability operations, we developed three interrelated analytic processes (as depicted in Figure S.1):

- Baseline activity analysis.
- Detailed activity assessment.
- Partner-selection modeling and exploratory analysis.

Without these analytic processes, security cooperation planners and programmers in the Army and other parts of DoD will be left to develop a BPC for stability operations strategy based solely on anecdotal information and personal opinions—as opposed to detailed, multifaceted, longitudinal data that have been systematically collected, aggregated, and interpreted for decisionmaking purposes.

Goals

In recent years, key U.S. government agencies have come to an agreement on the major goals for stability operations. The *Post-Conflict Reconstruction Essential Tasks* list produced by DOS[6] is organized into five broad technical sectors, which are quite similar to DoD's six major mission elements of a stability operation, found in DoD's *Military Support to Stabilization, Security, Transition, and Reconstruction Operations Joint Operating Concept.* Based on this DoD and DOS guidance, the fundamental goals of stability operations are to accomplish the following:

- Establish and maintain a safe and secure environment.
- Conduct strategic communications.
- Establish representative, effective governance and the rule of law.
- Deliver humanitarian assistance.
- Reconstruct critical infrastructure and restore essential services.
- Support economic development.

Largely absent from the existing documentation is an operational context to help decisionmakers prioritize and implement goals in a variety of pre-conflict, conflict, and post-conflict circumstances. Furthermore, DoD planning guidance, such as the BPC Execution Roadmap, establishes only a general connection between stability operations goals and BPC activities.[7]

Roles, Missions, and Capabilities

Although both concepts have deep historical roots, building partner capacity and stability operations have only recently migrated to positions near the top of the U.S. national security agenda. Furthermore, government officials have tended to consider the two topics separately rather than focus on the nexus between them. As a result, there is no

---

[6] This document is also referred to as the Essential Tasks Matrix.

[7] See Jennifer D.P. Moroney et al., *Building Partner Capabilities for Coalition Operations,* Santa Monica, CA: RAND Corporation, MG-635-A, 2007.

clearly defined and well-integrated strategy for using BPC activities to build stability operations capabilities in partner nations. In addition, key agencies have yet to reach a consensus on their respective roles and missions.

Mechanisms for aligning Army, DoD, and national BPC for stability operations strategy, planning, and resourcing should be constructed. Ideally, overall security sector assistance would be jointly managed by the departments of State and Defense. This could result in interagency objectives for employing and developing both departments' resources and capabilities for building partner capacity, as well as standardized procedures for formulating detailed BPC "roadmaps" for priority partners.

As part of a U.S. government BPC for stability operations strategy, the military should focus on its core security-related competencies; help civilian partners increase their operational capacities; tap into partner expertise when its own capacity is lacking, and reinforce the work of others—or simply get out of the way—when those others are doing a good job. Specifically, DoD's new emphasis on working "by, with, and through" partners requires further developing the list of essential BPC for stability operations capabilities; distinguishing between "direct" provision of stability operations assistance and BPC for stability operations; accounting for specialized stability operations activities as well as generic activities that could be useful for stability operations; and considering a range of BPC for stability operations contexts when selecting and prioritizing potential partners.

Baseline Activities Analysis

After sifting through U.S. government guidance in order to identify roles, missions, and capabilities for building partner capacity to conduct stability operations, the Arroyo team next examined the BPC for stability operations activities and programs currently being conducted by DoD, other U.S. government agencies, and major U.S. allies. This analysis aimed to help Army leaders better understand what BPC for stability operations programs and activities are now being conducted—both within the Army and elsewhere.

The U.S. Army has policy, planning, and resource management authority over only a small fraction of BPC for stability operations activities; most are controlled and managed by other DoD components, other U.S. government agencies (particularly DOS), and major U.S. allies. The lack of accessible, comprehensive data makes analysis of event-level BPC for stability operations difficult. However, in certain combatant commands (COCOMs), such as U.S. Southern Command, a significant number of events, resources, and personnel are focused on BPC for stability operations. Although both the United States and its allies focus on the security dimension of stability operations, the allies tend to have a longer-term investment approach to working with partners, primarily because of their cultural and colonial ties with certain countries and regions.

The Army should improve its visibility into security cooperation activities relevant to BPC for stability operations. In addition, the Army should design its security cooperation database so that it is not only interoperable with similar information systems across DoD, but also flexible enough to be used for analytical and operational purposes. Once it has acquired an overall understanding of ongoing BPC for stability operations activities, the Army should:

- Increase the number and extent of its BPC for stability operations activities.
- Expand its BPC for stability operations support in certain regions, such as U.S. Africa Command, where its programs are relatively scarce but where arguably the demand is growing.
- Re-evaluate its methods of delivering stability operations assistance to various partners—e.g., direct U.S. help or BPC aid, specialized stability operations activities, or general-purpose activities that could serve as building blocks for stability operations.
- Coordinate its BPC for stability operations efforts with those of its allies in order to reinforce and build upon their achievements as well as to direct limited U.S. resources to areas not currently receiving assistance.

- Make a concerted effort to learn from the BPC for stability operations experience of its allies, in particular, the United Kingdom and France, in several key areas such as trainer selection, mode of deployment, training of the trainers, and career implications for the trainer.

Detailed Activities Assessment

Building on the roles, missions, and capabilities synthesis and the baseline programmatic analysis described above, we next conducted an in-depth analysis of a range of BPC for stability operations programs. At the heart of this analysis is a six-step assessment approach designed to enable the Army and other DoD agencies to make more informed decisions about BPC for stability operations planning, programming, and budgeting (see Figure S.2). This approach provides a systematic method to evaluate existing security cooperation program and activity performance and effectiveness with respect to stability-related objectives and end states in particular countries.

Based on our analysis using the six-step approach, we found that BPC for stability operations activities tend to be more effective when they are used in the following ways:

- Applied in coordination with other, related activities to reinforce key concepts.
- Worked with, by, and through existing regional organizations and arrangements.
- Not "handed over" to an ally with little to no U.S. oversight.
- Sustained through careful planning and realistic resource allocation.

There tends to be greater follow-through (i.e., better outcomes) in an indigenous as opposed to a coalition operational context. Indigenous partners appeared genuinely interested in stability operations, especially disaster preparedness and response. Building partner capacity for coalition operations was more problematic given the political nature of out-of-country deployments.

Figure S.2
Six-Step Approach to Assess the Effectiveness of BPC for Stability
Operations

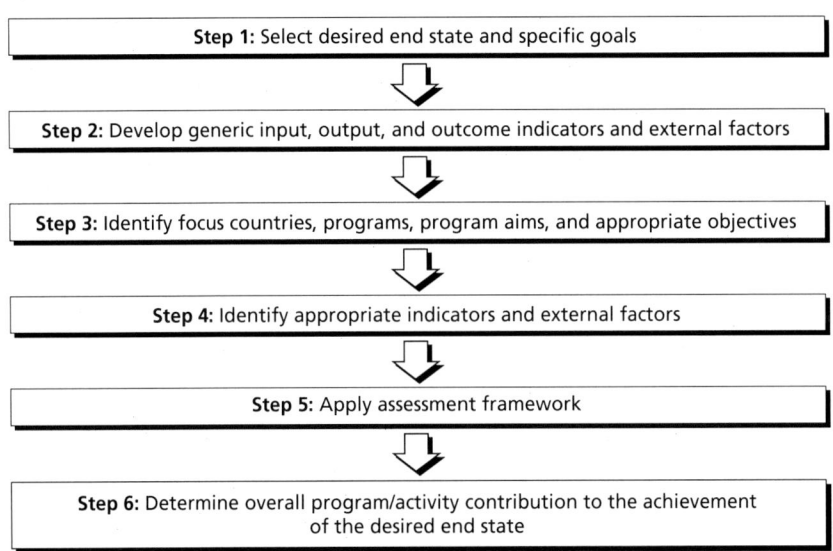

Step 1: Select desired end state and specific goals

Step 2: Develop generic input, output, and outcome indicators and external factors

Step 3: Identify focus countries, programs, program aims, and appropriate objectives

Step 4: Identify appropriate indicators and external factors

Step 5: Apply assessment framework

Step 6: Determine overall program/activity contribution to the achievement of the desired end state

RAND *MG942-S.2*

The Army should assist the COCOMs in developing a holistic approach to BPC for stability operations that

- is planned and resourced over a period of several years;
- involves all relevant U.S. military and civilian agencies and allies;
- targets multiple countries throughout a region; and
- employs a variety of security cooperation "tools" that are packaged and sequenced for each partner country.

The Army and DoD should consider the indigenous requirements of partners when designing BPC for stability operations activities and regional strategies. This may reduce the need for direct U.S. military assistance and increase the incentive for partners to engage in future coalition operations with the United States.

Analysis of Potential Partners

In an effort to provide some analytical rigor and standardization to the partner-selection approach, the Arroyo team developed a relatively simple spreadsheet method to help determine potential partners, assess the pros and cons of each partner, and choose ways to weight and assess selection factors.

Our exploratory analysis focused on identifying three types of potential stability operations partners:

- **Coalition partner.** A willing provider of significant stability operations-related capability in support of coalition operations outside the nation's own borders. A preferred partner demonstrates a moderate level of internal stability, international legitimacy, and strategic affinity with the United States.
- **Regional leader.** An actual or potential provider of capability and leadership for regionally based stability operations that are compatible with U.S. interests. Core regional partners demonstrate a moderate level of internal stability, international legitimacy, and strategic affinity with the United States.
- **Indigenous partner.** A fragile state, preferably receptive to U.S. government assistance and advice, whose deterioration or collapse could pose a significant threat to U.S. interests.

After introducing two partner-selection models—the regional/coalition model and the indigenous model—the Arroyo team used the models to conduct an exploratory partner analysis, which produced a detailed, country-by-country examination of potential BPC for stability operations partnerships. One of the most striking results of our exploratory analysis was that there are only a few well-rounded coalition and regional BPC for stability operations partners that are neither major allies nor advanced industrial states. That said, the number of potentially "willing" partners expands significantly if one values past participation in U.N. operations over involvement in U.S.-led operations. With respect to potential indigenous partners, domestic fragility and a lack of receptivity to outside intervention tend to go together.

In addition, finding strategically valuable indigenous partners that are receptive to U.S. help is especially difficult in the Middle East.

RAND Arroyo Center's exploratory partner analysis using the regional/coalition and indigenous models could support divergent courses of action. The apparent scarcity of high-potential partner nations could justify a narrowing of U.S. government BPC for stability operations efforts or serve as an impetus for greatly increasing the amount of resources dedicated to those efforts. Less ambiguously, the United States should consider focusing more on coalition and regional candidates with a demonstrated willingness to participate in U.N. deployments. Because few countries are both fragile and receptive, the decision to attempt to build indigenous stability operations capacity may, in many cases, have to be based on the degree of a country's internal weakness and the salience of the U.S. strategic interest in that country.

Ideally, the results of these analytical processes will have a significant effect on the set of BPC for stability operations activities and partners, aligning relevant and effective activities with appropriate partners.

Acknowledgments

The authors owe a great debt to a number of officers, civil servants, and analysts for their assistance on this study. These include current and past members of the Office of the HQDA G-35, Director of Strategy, Plans and Policy, especially the Multinational Strategy and Programs Division and the Army Stability Operations Division; the National Guard Bureau, Office of International Affairs; U.S. Southern Command Theater Security Cooperation Office, J-5; Southern Command State Partnership Program; U.S. European Command Theater Security Cooperation office, J-5; the State Department's African Affairs Bureau; the Office of the Secretary of Defense for Policy (African Affairs); the Peacekeeping and Stability Operations Center, Army War College, Carlisle, PA; the George C. Marshall Center, Garmisch, Germany; the Center for Hemispheric Defense Studies and the Near East South Asia Center for Strategic Studies, National Defense University; the Asia Pacific Center for Security Studies, Honolulu, HI; U.S. Institute for Peace; the Office of the Secretary of Defense for Policy/Partnership Strategy; and the U.S. Defense Attaché Office in Rome, Italy.

At RAND we also wish to acknowledge the outstanding editorial and formatting support of Hilary Wentworth.

The project officers for this study were Mr. Mark McDonough, Chief of the Multinational Force Compatibility Branch in Army Staff G-35, and Mr. Hartmut Lau, Chief, Policy, Plans, and Assessments Branch. Mr. McDonough and Mr. Lau provided outstanding support to the study on both substantive and administrative matters. We are grateful for their guidance and help throughout this one-year effort.

List of Abbreviations

ACOTA	African Contingency Operations Training and Assistance
ACRI	African Contingency Response Initiative
AFRICOM	United States Africa Command
AMIS	African Union Mission in Sudan
AOR	Area of Responsibility
APCSS	Asia-Pacific Center for Strategic Studies
ARGOS	Army Global Outlook System
BPC	Building Partner Capacity
CENTCOM	United States Central Command
CHDS	Center for Hemispheric Defense Studies
CJCSM	Chairman of the Joint Chiefs of Staff Manual
CMEP	Civil-Military Emergency Preparedness
COCOM	Combatant Command
COE-DAT	Center for Excellence Against Terrorism
CoESPU	Italian Center of Excellence for Stability Police Units
DHS	Department of Homeland Security
DOC	Department of Commerce
DoD	Department of Defense
DoDD	Department of Defense Directive
DOE	Department of Energy
DOJ	Department of Justice
DOS	Department of State

DOT	Department of Transportation
DPKO	Department of Peacekeeping Operations
EIU	Economist Intelligence Unit
EPS	Environmental Protection Agency
ETM	Essential Tasks Matrix
EUCOM	United States European Command
EXBS	Export Control and Related Border Security Assistance
FHA	Federal Housing Administration
FM	Army Field Manual
FY	Fiscal Year
GCMC	George C. Marshall Center
GDP	Gross Domestic Product
GIES	Romanian Ministry of General Inspectorate for Emergency Situations
GIS	Geospatial Information System
GPOI	Global Peace Operations Initiative
HQDA	Headquarters, Department of the Army
HQDA G-35 SSO	Headquarters, Department of the Army, G-35, Stability Operations Division
ICITAP	International Criminal Investigative Training Assistance Program
ITA	International Trade Administration
JCS	Joint Chiefs of Staff
JOC	Joint Operating Concept
JP	Joint Publication
LLO	Logical Line of Operation
MME	Major Mission Element
MPEP	Military Personnel Exchange Program
NDU	National Defense University
NESA	Near East South Asia Center for Strategic Studies
NGO	Nongovernmental Organization

NOAA	National Oceanic and Atmospheric Administration
NSPD	National Security Presidential Directive
NTIA	National Telecommunications and Information Administration
OSD	Office of the Secretary of Defense
PfP	Partnership for Peace
PKO	Peacekeeping Operations
S/CRS	Department of State Coordinator for Reconstruction and Stabilization
SPP	State Partnership Program
SOUTHCOM	United States Southern Command
SSTR	Stability, Security, Transition, and Reconstruction
TSCMIS	Theater Security Cooperation Management Information Systems
TTX	Table Top Exercise
U.N.	United Nations
USAID	United States Agency for International Development

CHAPTER ONE

Introduction

Secretary of Defense Robert M. Gates has defined the current strategic preoccupation of the U.S. military, the war on terror, as "a prolonged, worldwide irregular campaign—a struggle between the forces of violent extremism and those of moderation." In order to effectively carry out such a campaign, Gates writes, the military must learn two hard lessons from the wars it has conducted in Afghanistan and Iraq since the fall of 2001.

The first lesson is that "over the long term, the United States cannot kill or capture its way to victory."[1] In other words, "soft power"—including diplomacy, strategic communications, foreign assistance, civic action, and economic reconstruction—is at least as important as, if not more than, "hard power" in creating the conditions for the eventual defeat of violent extremism throughout the world.[2] Thus, on the one hand, Gates has called for the U.S. government to greatly increase the capacity of civilian national security agencies—in particular, the U.S. Department of State (DOS) and the U.S. Agency for International Development (USAID)—so they can take the lead in exercising soft power in unstable parts of the globe. On the other hand, the Defense Secretary has recognized that the military must continue to play a major role in stability operations: maintaining security, providing humanitarian aid, beginning reconstruction, bolstering local gov-

---

[1] Robert M. Gates, "A Balanced Strategy: Reprogramming the Pentagon for a New Age," *Foreign Affairs*, January/February 2009.

[2] "U.S. Defense Chief Urges Greater Use of 'Soft Power,' Agence France-Presse, November 26, 2007.

ernments and public services, especially "in the midst of or in the aftermath of conflict."[3]

Gates's second major strategic lesson from recent U.S. interventions in the Middle East is the desirability of taking an "indirect approach" to prosecuting the war on terror. In his view, the United States is unlikely to mount another invasion and occupation on the scale of that undertaken in Afghanistan or Iraq in the foreseeable future. To the extent possible, the United States should follow a sustainable counterterrorism strategy, one that does not rely on the massive application of U.S. combat power. Ideally, the United States should work "by, with, and through" its allies and partners and, when necessary, bolster the capacity of their governments and security forces to effectively contribute to the war on terror. According to Michael G. Vickers, former Assistant Secretary of Defense for Special Operations/Low-Intensity Conflict and Independent Capabilities, the Pentagon envisions "a global network . . . made up of the U.S. and foreign militaries and other government personnel in scores of countries with which the United States is not at war . . . designed to wage 'steady state' counterterrorism operations."[4]

Given the Secretary's guidance to focus on bolstering the ability of our allies and partners to effectively contribute to the war on terror, the Army must develop a strategy to fund, develop, and execute security cooperation programs and activities that help build other countries' capacity to conduct stability operations—the key operational component in most large-scale counterterrorism/insurgency campaigns. Such programs and activities need to be appropriately packaged and associated with particular stability operations capabilities. They also need to be reconciled with similar capacity-building programs that are being undertaken by other Department of Defense (DoD) organizations, other U.S. government agencies, and major U.S. allies. Ideally,

---

[3] Gates.

[4] Ann Scott Tyson, "U.S. To Raise 'Irregular War' Capabilities," *Washington Post*, December 4, 2008, p. 4.

the conduct of Army programs should be informed by assessments of what has worked in the past in terms of building stability operations capacity. Finally, limited Army capacity-building resources should be directed to partner countries in accordance with U.S. strategic priorities and objectives outlined in such documents as the Guidance for the Employment of the Force, regional combatant command (COCOM) campaign plans, and the Army Security Cooperation Plan.

This report, which provides the results of a RAND Arroyo Center study for the U.S. Army International Affairs Office, addresses some of the practical issues in designing and implementing an Army strategy for helping selected allies and partners to build stability operations capacity. Specific issues addressed in this report include:

- What are the doctrinal components of building partner capacity (BPC) for stability operations?
- What kinds of security cooperation programs are being conducted by the United States and its major allies that relate to BPC for stability operations?
- How might the government assess the effectiveness of these programs?
- What kinds of partner countries might be appropriate recipients of future stability operations assistance?

This report is not intended to be a comprehensive review of previous and ongoing efforts by the U.S. military and others to build the capacity of foreign countries to conduct stability operations. Nor does this report provide a detailed step-by-step roadmap for building stability operations capacity in specific countries. Rather, this study presents a conceptual and analytical framework for the Army to understand the key elements of building stability operations capacity and develop methods for assessing BPC for stability operations activities and potential partners.

Understanding Building Partner Capacity and Stability Operations

An important objective of our study is to clarify the relationship between BPC and stability operations. At present, there is no strategic guidance for BPC for stability operations per se; there is BPC guidance, and there is stability operations guidance.

Building Partner Capacity

Building partner capacity is a new name for a diverse set of governmental activities that have recently lacked strategic coherence. During the Cold War era, the rationale for U.S. foreign assistance programs was clear: to bolster the defenses of pro-Western regimes confronted either by an external military threat from the Soviet Union and other communist powers or an internal threat from communist-supported insurgents. In line with the containment doctrine, successive U.S. administrations were more or less successful in their efforts to integrate the diplomatic, military, economic, and informational aspects of power for the purpose of increasing the capabilities of U.S. allies and partners as part of the Cold War.

In the post–Cold War era, the strategic rationale for U.S. foreign assistance became less clear, and the interagency framework, which had coordinated the many facets of U.S. foreign policy and national security, fragmented into often competing departmental factions. Consequently, the U.S. government has largely dealt with each post–Cold War international crisis only after it has emerged, and then in an ad hoc, initially uncoordinated fashion that has treated allies and partners mostly as an afterthought. Exceptions to this rule, such as the interagency initiative to train, equip, and sustain the Republic of Georgia's security forces for duty along its borders with Chechnya and in Iraq, have not been part of an overall BPC strategy that balances partner requirements with U.S. interests.

As a result of the September 11, 2001 attacks by Al Qaeda and the subsequent difficulties the U.S. military has encountered combating Islamist insurgencies in Iraq and Afghanistan, DoD has begun to think more strategically about BPC as a means of addressing emerging global threats to the United States and its allies. For example, the 2006 Quadrennial Defense Review and the 2006 BPC Execution Roadmap emphasize the importance of improving the security and defense capabilities of partner countries for coalition and indigenous operations.[5] Additionally, the position of Assistant Secretary of Defense for Global Affairs was created in 2006 to reflect this new focus on building partner capacity.[6]

As currently conceived, BPC is a multi-agency, multinational initiative that draws on the elements of security cooperation[7] to achieve U.S. strategic objectives that include

- defeating terrorist networks;
- preventing hostile states and nonstate actors from acquiring or using WMD;
- conducting irregular warfare and stability operations; and
- enabling host countries to provide good governance.[8]

---

[5] What we are calling for brevity the "BPC Execution Roadmap" is the *Quadrennial Defense Review Building Partnership Capacity Execution Roadmap,* Office of the Secretary of Defense and the Joint Staff J-5, Washington, D.C., May 2006; it is an evolving concept. It not only includes guidance on how DoD should train and equip foreign military forces, but also discusses the need to improve the capacity of other security services within partner countries. Moreover, the concept also refers to the need to improve DoD's ability to work with nonmilitary forces in an operational context for integrated operations.

[6] Under this assistant secretary of defense, two new offices were created: "Partnership Strategy" and "Coalition Affairs," each headed by a deputy assistant secretary of defense. A third office called "Global Threats" was also added, which combined counterproliferation, counternarcotics, and transnational threats.

[7] According to the Defense Security Cooperation Agency (DSCA) web site, security cooperation includes "those activities conducted with allies and friendly nations to: build relationships that promote specified U.S. interests, build allied and friendly nation capabilities for self-defense and coalition operations, [and] provide U.S. forces with peacetime and contingency access." Security assistance is a subset of security cooperation and consists of "a group of programs, authorized by law that allows the transfer of military articles and services to friendly foreign governments." These programs include Foreign Military Sales, Foreign Military Financing, and International Military Education and Training. See DSCA's web site's FAQ section, and the *Security Assistance Management Manual,* DoD 5105.38-M, 2007 (also available online). A full listing of security assistance programs may be found on p. 33 of the manual.

[8] BPC Execution Roadmap, Washington, D.C., May 2006.

The role of U.S. allies and partners in this effort is both as a force multiplier and as a hedge against future security requirements. With greater global demands for U.S. forces and an expanding list of adversaries, conditions, and crises that could threaten U.S. national interests, allies and partners increase and diversify the capabilities needed to fight in future conflicts, particularly on unfamiliar geographical and cultural terrain.[9]

Stability Operations

Stability operations is an evolving and variously named concept.[10] Historically, the U.S. military tended to relegate operations that did not involve full-scale combat to several overlapping but not identical categories: small wars; low-intensity conflicts; military operations other than war; small-scale contingencies; peace operations; stability and support operations; stability, security, transition, and reconstruction (SSTR) operations; or simply, stability operations.[11] Stability operations, as defined by DoD, are military missions, tasks, or activities conducted in foreign countries and in coordination with other instruments of national power to maintain or re-establish a safe and secure environment, provide essential government services, reconstruct emergency infrastructure, and deliver humanitarian relief.[12]

---

[9]   White House, *The National Security Strategy,* September 2002, p. 29.

[10]   As indicated earlier, the bulk of the research in this report on stability operations was completed in early to mid-2007, when the concept of stability operations remained in flux. Since then, the U.S. Army has published two key field manuals discussing stability operations, FM 3-0, *Operations,* in February 2008, and FM 3-07, *Stability Operations,* in October 2008. Also, DoD published Joint Publication (JP) 3-07.3, *Peace Operations,* in October 2007, and Instruction 3000.05, *Stability Operations,* September 16, 2009. These new documents largely formalized earlier discussions about stability operations, and what is written here is consistent with these later documents.

[11]   DoD defines stability operations as "military and civilian activities conducted across the spectrum from peace to conflict to establish or maintain order in States and regions." See U.S. Army, *Stability Operations,* FM 3-07, 2003 and 2008; Department of Defense Directive 3000.05, *Military Support for Stability, Security, Transition, and Reconstruction Operations,* November 28, 2005 (hereafter, DoDD 3000.05, *Military Support for SSTR Operations*); and Department of Defense Instruction 3000.05, *Stability Operations,* September 16, 2009.

[12]   Joint Chiefs of Staff (JCS), *Joint Operations,* JP 3-0, February 2008, p. V-1.

Despite their differences, all of these concepts refer to military operations in civilian environments. Stability and support operations are typically interagency and often multilateral. Support operations, i.e., humanitarian or environmental assistance, are "civil-military" by definition. A recently published U.S. Army field manual, *Stability Operations*, includes an array of missions, among them peace operations, combating terrorism, counterdrug operations, population control, and nation assistance.[13]

The joint military community, however, continues to use a more cumbersome term that largely overlaps with the Army's definition of stability operations: military support for stabilization, security, reconstruction and transition operations.[14] In the interest of brevity, we use the Army's "stability operations" term throughout this report.

From the mid-1970s to the early 1990s, DoD's focus on major combat operations meant that stability operations were relatively neglected. Despite a long and continuous record of U.S. military involvement in operations other than war, stability operations were considered "lesser included cases" when it came to force planning. Troops that were trained and equipped for combat were assumed to be capable of performing noncombat missions, such as humanitarian assistance, disaster relief, and peacekeeping, without much additional guidance. Furthermore, following the U.S. defeat in Vietnam, DoD leaders consciously eschewed tasks that smacked of nation building, believing that the military should stick to its core competency of warfighting. Most DoD leaders believed political and economic development was the responsibility of the State Department, USAID, multilateral and nongovernmental aid organizations, and the private sector, not DoD.

Nevertheless, unforeseen international events trumped the military's desire to emphasize large-scale combat operations. U.S. interventions in Somalia, Haiti, and the Balkans in the 1990s, and more

---

[13] Headquarters, Department of the Army (HQDA), *Stability Operations*, FM 3-07, October 2008.

[14] HQDA, *Operations,* FM 3-0, February 2008; and DoDD 3000.05, *Military Support for SSTR Operations.*

recently in Afghanistan and Iraq, dramatically raised the profile of stability operations. They also significantly eroded the distinction between military and civilian responsibilities in a conflict environment. With its enormous resources, operational orientation, and inherent security capacity, DoD undertook stabilization and reconstruction tasks that civilian agencies could not, or would not, perform.

After the war on terror began in the wake of 9/11, many Bush administration officials became proponents of stability operations (as evidenced in several national security documents, which are detailed in Chapter Two). President Bush's National Security Presidential Directive (NSPD) 44, issued in 2005, made stabilization and reconstruction of conflict-torn countries and regions an essential part of the war on terror, declaring it the United States' responsibility to help these places "establish a sustainable path toward peaceful societies, democracies, and market economies."[15] The State Department's newly established Coordinator for Reconstruction and Stabilization (S/CRS) was given the mission to "lead, coordinate and institutionalize" civilian government agencies' support for this initiative.[16] For its part, in 2005 DoD elevated stability operations to a "core U.S. military mission" comparable in priority to combat operations, which needed to be "integrated across all DoD activities including doctrine, organizations, training, education, exercises, materiel, leadership, personnel, facilities, and planning."[17] Although DoD recognized that it would undertake stability operations "in support of a broader U.S. Government effort," the 2006 Joint Operating Concept (JOC) for stability, security, transition, and reconstruction acknowledged that the military might become involved in a range of noncombat tasks, to include restoring essential services, promoting economic development, and administering occu-

---

[15] The White House, National Security Presidential Directive 44, *Management of Interagency Efforts Concerning Reconstruction and Stabilization,* December 7, 2005 (hereafter, NSPD-44).

[16] Department of State, Office of the Coordinator for Reconstruction and Stabilization web site.

[17] DoDD 3000.05, *Military Support for SSTR Operations.*

pied territory.[18] In addition, in 2008 the Army issued an updated edition of FM 3-0, *Operations,* which defines stability operations as one of the three basic components in the Army full spectrum of operations, along with offensive and defensive operations.[19]

Although there is clear strategic guidance on the stability operations goals, largely absent from the existing documentation is an operational context to help decisionmakers prioritize and implement goals in a variety of pre-conflict, conflict, and post-conflict circumstances. Furthermore, DoD planning guidance, such as the BPC Execution Roadmap, establishes only a general connection between stability operations goals and BPC activities.[20]

Analytical Approach: Strategic Elements of Building Partner Capacity for Stability Operations

In this study, we limited our investigation of building partner capacity for stability operations to situations that do not involve large numbers of U.S. military forces. Instead, this study focused on small-scale capacity-building activities during periods of relative peace. This excludes consideration of large-scale security force assistance to Iraq and Afghanistan. This is not because such assistance is unimportant or unworthy of examination from a capacity-building perspective. On the contrary, there is already a great deal of policy and academic emphasis on the impact of U.S. and coalition efforts to train, assist, and advise Iraqi and Afghani security forces. However, there is considerably less analytic focus on building partner capacity in the rest of the world, even though the rest of the world arguably provides a better laboratory for analyzing how relatively small amounts of foreign assistance might

---

[18] Department of Defense, Joint Forces Command, *Military Support to Stabilization, Security, Transition, and Reconstruction Operations Joint Operating Concept,* December 2006 (hereafter DoD, *Military Support to SSTR Operations JOC*).

[19] HQDA, *Operations, FM 3-0,* February 2008.

[20] See Jennifer D.P. Moroney et al., *Building Partner Capabilities for Coalition Operations,* Santa Monica, CA: RAND Corporation, MG-635-A, 2007.

facilitate the Secretary of Defense's vision of a shared responsibility among the United States and its partners for "heading off the next insurgency" or "stopping the collapse of the next failed state."[21]

That said, we have conducted an exploratory analysis of five strategic elements necessary to align U.S. government security cooperation efforts with the goal of building partner capacity for stability operations in a largely peacetime environment. Figure 1.1 lays out the organization of the strategic elements as well as the corresponding chapter in which each element is examined in depth. Below we introduce each of these options.

Goals

In recent years, key U.S. government agencies—particularly the Department of State and DoD—have come to an agreement on major stability operations goals. The document *Post-Conflict Reconstruction Essential Tasks* produced by S/CRS[22] is organized into five broad technical sectors, which have been adopted directly by emerging joint military peace operations doctrine.[23] They are also quite similar to DoD's six major mission elements (MMEs) of a stability operation, found in the *Military Support to Stabilization, Security, Transition, and Reconstruction Operations Joint Operating Concept,* with the major difference being that DOS incorporates strategic communications within each of its sectors, whereas the DoD JOC treats strategic communications as a separate (albeit cross-cutting) MME.[24] Based on DoD and DOS guidance, the fundamental goals of stability operations are to accomplish the following:

---

[21] Robert Burns, "Gates: Extremist Threat Requires New U.S. Approaches," *Washingtonpost.com,* October 15, 2008.

[22] This document is also referred to as the Essential Tasks Matrix. See Department of State, Office of the Coordinator for Reconstruction and Stabilization, *Post-Conflict Reconstruction Essential Tasks,* April 2005. As of January 2010:
http://www.crs.state.gov/index.cfm?fuseaction=public.display&shortcut=J7R3

[23] DoD, JP 3-07.3, *Peace Operations,* 2007.

[24] Major mission element and logical line of operation are synonymous terms. DoD, *Military Support to SSTR Operations JOC,* December 2006, p. 20.

- Establish and maintain a safe and secure environment.
- Conduct strategic communications.
- Establish representative, effective governance and the rule of law.
- Deliver humanitarian assistance.
- Reconstruct critical infrastructure and restore essential services.
- Support economic development.

Roles and Missions

Although there is improved clarity in the U.S. government about BPC for stability operations goals, there is still considerable uncertainty about key BPC for stability operations roles and missions, particularly among the State Department and DoD players (see Chapter Two).

Officially, the Department of State has been assigned the lead role in coordinating overseas reconstruction and stabilization, including building partnership capacity in these areas.[25] That said, DOS has primarily focused on planning for post-conflict reconstruction following a U.S.-organized intervention, while its subordinate organization,

Figure 1.1
BPC for Stability Operations: Strategic Elements

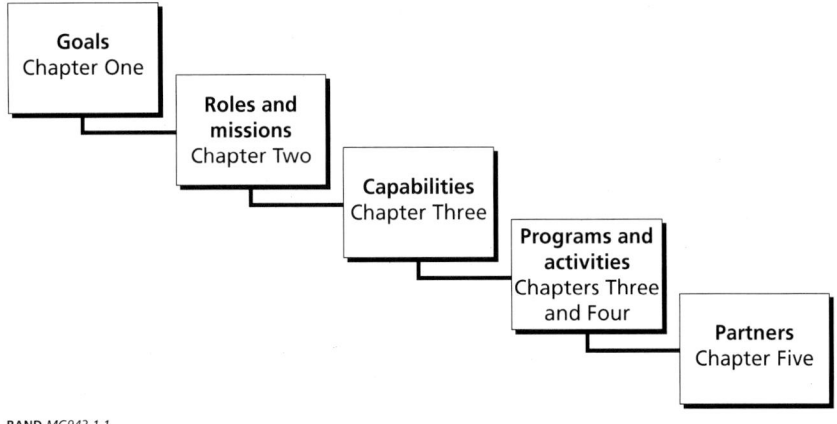

RAND MG942-1.1

---

[25] NSPD-44, p. 4.

USAID, has continued in its traditional role of providing long-term development assistance to failing or unstable states.

In reality, the DOS and other civilian departments currently lack the *capacity*—personnel, funding, security forces, and operational know-how—to manage large-scale capacity-building activities in a conflict environment.

On the other hand, as the results of U.S. counterinsurgency operations in Afghanistan and Iraq have made clear, DoD lacks much of the *capability*—technical, cultural, linguistic, legal, financial, political, etc.—to help rebuild a shattered society. While civil-military Provincial Reconstruction Teams and Joint Interagency Coordinating Groups show some promise of resolving the roles-and-missions issue at the tactical and operational levels, these organizations are in their nascent stages and are only beginning to work together at the strategic level. Finally, the United States and its major allies have yet to begin a serious dialogue on ways to coordinate their respective foreign assistance programs.

Capabilities

Both DoD and the State Department have produced BPC and stability operations doctrine that we have used to help organize the types of capabilities that the Army needs in order to conduct a wide range of activities in support of broad stability goals.

The State Department has established a reasonably comprehensive list of tasks and subtasks for use in post-conflict stability planning. However, because this list, called the Essential Tasks Matrix (ETM), is too skeletal a framework to form the basis for a capacity-building strategy, we have constructed working definitions of executable "capabilities" using a combination of the ETM and several DoD and DOS doctrinal sources. For example, the U.S. Army's peace operations and stability operations field manuals and the State Department's *Foreign Assistance Standardized Program Structure and Definitions*[26] each con-

---

[26] JCS, *Universal Joint Task List,* CJCSM 3500.04D, August 2005; JCS, *Peace Operations,* JP 3-07.3, October 2007; DOS, *Foreign Assistance Standardized Program Structure and Definitions,* 20 October 2006; ALSA Center, *Multi-Service Tactics, Techniques and Procedures for Conducting Peace Operations,* FM 3-07.31, October 2003; and HQDA, *Stability Operations,* FM 3-07, 2008.

tain information that is helpful for clarifying the capabilities necessary to establish a safe and secure environment. These capabilities include conducting peace operations, developing and sustaining armed services and intelligence services, and establishing and maintaining boundary control. (See Chapter Two and Appendix A for a detailed description of proposed stability operations capabilities).

Programs and Activities

The departments of Defense and State, USAID, other U.S. government agencies, and key U.S. allies, such as the United Kingdom, France, and Italy, all conduct peacetime security cooperation programs and activities that, to a greater or lesser extent, contribute to building the capacity for stability operations (see Chapter Three).

"Programs" and "activities" are used here as catchall terms for government-supported security cooperation initiatives, programs, activities, and events. Technically, an initiative is a coordinated grouping of usually interagency programs and activities that are often directed at a single country or region, such as Plan Colombia. Programs are activities that are specifically funded in the budget, such as Foreign Military Financing or International Military Education and Training. Generic activities are not specifically funded; they include interactions between the services and foreign military and civilian government officials that are funded via operating accounts. Events are discrete examples of programs or activities, such as a military training team marksmanship class, a U.S. Navy port visit, or a meeting of international scientific and technical experts.

These programs and activities encompass a variety of security cooperation methods or "ways," including conferences, workshops and information exchanges, defense and military contacts, education, equipment and infrastructure, exercises, and training. They also support all six of the stability operations goals, detailed above.

Partners

As the United States places greater emphasis on BPC and stability operations, it should take a more analytical approach to evaluate and select potential capacity-building partners for different kinds of stability operations. In the 1990s, the focus of U.S. capacity-building efforts was on its major treaty allies, particularly those in NATO, whose collective defense spending had been in steep decline since the end of the Cold War. With the United States' growing reliance on "coalitions of the willing" in the war on terror, the DOS and DoD have begun to identify and cultivate countries that are new allies or non-ally partners, especially those that can contribute certain "niche capabilities."[27] More recently, U.S. military overstretch—combined with a concern about weak states as possible breeding grounds for international terrorists and other criminals and an increasing emphasis on irregular warfare—has created a need to further expand the boundaries of capacity building to include different types of partners.

In this study, we analyze three general types of stability operations partners, with the focus being on countries that are neither major allies nor advanced industrial states. A *coalition partner* is an actual or potential provider of significant stability-related capability in support of U.S.-led coalition operations outside its borders. A *regional leader* is an actual or potential provider of leadership and capability for regionally based stability operations compatible with U.S. interests. In both cases, the most appropriate partners demonstrate a moderate level of internal stability, international legitimacy, and strategic affinity with the United States. By contrast, an *indigenous partner* is a fragile state with a minimal capacity to use, and a willingness to accept, U.S. government assistance for internal stability operations. Preferred indigenous partners include those wherein the collapse of public security and authority would pose a significant threat to U.S. interests.

Analyses Useful for Building Partner Capacity Planning

In order to integrate the five elements essential to building partner capacity for stability operations, we have developed three interrelated analytic processes:

---

[27] See Jennifer D.P. Moroney et al., *A Capabilities-Based Strategy for Army Security Cooperation*, Santa Monica, CA: RAND Corporation, MG-563-A, 2007.

- baseline activity analysis;
- detailed activity assessment; and
- partner-selection modeling and exploratory analysis.

Without these analytic processes, security cooperation planners and programmers in the Army and other parts of DoD will be left to develop a BPC for stability operations strategy based solely on anecdotal information and personal opinions—as opposed to detailed, multifaceted, longitudinal data that has been systematically collected, aggregated, and interpreted for decisionmaking purposes. Figure 1.2 builds on Figure 1.1 by including these analyses.

Baseline Activities Analysis
The first step in developing a realistic and executable BPC for stability operations plan is to conduct a baseline analysis of U.S. and major ally

Figure 1.2
BPC for Stability Operations: Strategic Elements with Related Analyses

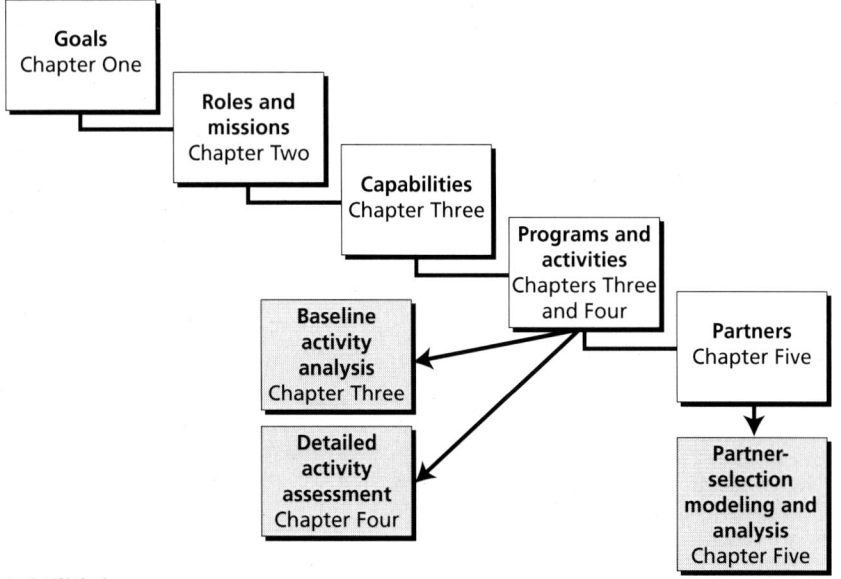

RAND *MG942-1.2*

security cooperation activities that pertain to building partner capacity for stability operations. This descriptive analysis would attempt to answer the following questions:

- What kinds of activities are currently being undertaken?
- By whom and with whom are they being conducted?
- Toward what ends are these activities being pursued?
- What is the extent of the resources that are being expended in terms of money and personnel?
- What are the sources of funding for these activities?
- What gaps and overlaps exist in the programs of importance to building partner capacity for stability operations?

Unfortunately, answering these basic questions in a comprehensive way is very difficult. Although significant strides have been made in recent years at the regional combatant command level with the establishment of Theater Security Cooperation Management Information Systems (TSCMIS) and in the Army with the Army Global Outlook System (ARGOS), these databases are neither fully integrated with one another nor linked to other relevant DoD information systems. Furthermore, they do not contain data on many security cooperation programs managed by DoD and other government agencies. Finally, TSCMIS data are not collected in a consistent or thorough manner.

Recognizing the limited nature of the available data, we have attempted to demonstrate two kinds of baseline activity analysis. On the one hand, we have collected broad (albeit somewhat superficial) information on U.S. Army, other DoD, other U.S. agency, and major ally stability-related programs and activities. On the other hand, we have conducted an in-depth analysis of the U.S. Southern Command's (SOUTHCOM) TSCMIS for fiscal years (FYs) 2005–2006 in order to describe BPC for stability activities within SOUTHCOM's area of operation. In addition, we have attempted to differentiate between events that involve direct U.S. provision of stability benefits versus events designed to build partner capacity for stability operations.

Detailed Activities Assessment

The next analytical step in developing a BPC for stability operations strategy is to assess the performance and effectiveness of ongoing activities—the goal being to understand what works, in what way, and in what context. At present, however, aggregated, long-term, outcome-related security cooperation assessment data are even less available than disaggregated, short-term input, or output-related information. For the most part, security cooperation evaluations are generally subjective and based on changing and ill-defined measurement criteria that cannot be reliably combined to provide a regional, much less global, understanding of the effectiveness of capacity-building activities that relate to stability operations.

In an effort to help the Army fill this gap, we developed the following six-step case study approach to assess the effectiveness of stability-related security cooperation programs and activities in particular countries and operational contexts (see Chapter Four).[28]

1. Select cases representative of different ways to build partner capacity for stability operations.
2. Choose a stability operations goal for analysis that is of significant interest to DoD and the U.S. Army (i.e., establish a safe and secure environment).
3. Develop generic output and outcome indicators for various capacity-building ways (e.g., exercises)[29] and then select the indicators that applied to each specific case.
4. Collect documentary evidence and conduct focused discussions with U.S. government and partner country officials related to the chosen output and outcome indicators.
5. Investigate external factors that had a significant influence on activity effectiveness.

---

[28] This effort builds on a previous study undertaken by RAND Arroyo Center for HQDA G-35 in FYs 2003–2004: Jefferson P. Marquis et al., *Assessing the Value of U.S. Army International Activities*, Santa Monica, CA: RAND Corporation, MG-329-A, 2006.

[29] See Appendix C for a list of generic output and outcome indicators related to the stability operations objective of establishing a safe and secure environment.

6.   To the extent possible, assess the impact of each BPC case in achieving the stability operations goal in particular European, Latin American, and African countries.

After introducing the six-step approach, we then applied it to a number of relevant case studies, which are detailed in Chapter Four and Appendix D.

Partner-Selection Modeling and Analysis

The final phase of our analysis is designed to help the Army select and prioritize partner nations for capacity-building activities. Although there are a number of DOS, DoD, and Army guidance documents that discuss the selection of partner nations, the current process for prioritizing security cooperation partners lacks analytical rigor as well as consistency across agencies and over time. Furthermore, the process does not explicitly examine partners from different operational perspectives.

For this analysis, we developed two partner-selection models: one for coalition and regional partners, and one for indigenous partners (see Chapter Five). For each model, we developed key attributes and associated indicators that determined the ranking of most of the world's countries depending on the weighting of these attributes and indicators. Evidence for the indicators associated with the key attributes came from a variety of sources, including the World Bank, the United Nations, the U.S. State Department, the Fund for Peace–Failed States Index, and the Economist Intelligence Unit. Rather than using these models to develop a static list of prioritized partner countries, we conducted an exploratory analysis designed to demonstrate how their country and regional importance might change depending on how policymakers weigh various selection criteria.

Organization of the Report

As illustrated above in Figures 1.1 and 1.2, the remainder of the report is organized as follows: Chapter Two examines the roles and responsibilities associated with capacity building and stability operations, as well as the missions and capabilities required for carrying out these responsibilities. Chapter Three describes the universe of capacity-building activities related to stability operations from a global and regional perspective; identifies possible programmatic gaps and overlaps; and provides in-depth analysis of SOUTHCOM's TSCMIS for FYs 2005–2006 to describe BPC for stability activities within SOUTHCOM's area of operation by BPC method (also called "way"), stability operations goal/capability, and partner nation in terms of numbers of events, amount of authorized funding, and numbers of foreign participants. Chapter Four provides a six-step approach to assess the effectiveness of stability-related security cooperation programs and activities and applies that approach to detailed case studies. Chapter Five identifies the principal attributes of potential partners in different operational contexts (coalition, regional, and indigenous); develops measures, indicators, and models; and conducts exploratory analysis. Chapter Six provides our conclusions based on our five major elements of BPC planning—goals, roles and missions, capabilities, programs and activities, and partners—and makes recommendations to assist Army, other DoD organizations, and other U.S. government departments in developing and implementing a future approach to building partner capacity for stability operations.

Five appendixes follow the conclusion of the main part of the report. Appendix A provides working definitions for the stabilization capabilities identified in Chapter Two, based on the DOS Essential Tasks Matrix and DoD doctrine. To supplement the analysis in Chapter Three, Appendix B briefly describes the wide range of BPC programs related to stability operations currently being executed by the U.S. Army, other DoD organizations, other non-DoD government agencies, and major U.S. allies. Appendix C provides a list of generic indicators for measuring the performance and effectiveness of stability-related programs focused on achieving a safe and secure environment. Appendix D presents three capacity-building case studies from Chapter Four that illustrate our assessment methodology in greater detail. Appendix E provides a technical description of the coalition and indigenous models used to conduct the exploratory analysis of stability partners in Chapter Five.

CHAPTER TWO

BPC for Stability Operations: Roles, Missions, and Capabilities

Although both concepts have deep historical roots, building partner capacity and stability operations have only recently migrated to positions near the top of the U.S. national security agenda. Furthermore, government officials have tended to consider the two topics separately rather than focus on the nexus between them. As a result, there is no clearly defined and well-integrated strategy for using BPC activities to build stability operations capabilities in partner nations. In addition, key agencies have yet to reach a consensus on their respective roles and missions.

Until recently, BPC guidance has been directed toward facilitating U.S.-led coalition operations, giving less attention to the advantages of developing the capacity of indigenous and regional partners. While the State and Defense departments appear to agree on overall stability operations goals, operational capabilities have not been spelled out in enough detail for executive agents like the U.S. Army and the other services, which are tasked with implementing BPC for stability operations. In addition, BPC for stability operations activity categories are not sufficiently defined for assessment purposes, which impedes the Army's ability to evaluate the effectiveness of particular programs or activities.

This chapter is intended to help the Army address two key questions:

- What are the roles and missions associated with BPC and stability operations?
- What are the unique capabilities required for carrying out these responsibilities?

To help answer these questions, we first identified and reviewed key guidance documents that collectively provide broad direction for BPC roles. Next, we examined the guidance documents for the planning and conduct of stability operations, particularly at the DOS and DoD levels. Finally, we examined how the Army has integrated the higher-level BPC and stability operations guidance and incorporated the two into guidance that defines the roles, missions, and specific capabilities needed to conduct BPC for stability operations.

This chapter's methodical examination of BPC for stability operations roles, missions, and capabilities serves as the foundation for the following chapters that address BPC for stability operations activities, assessments, and partners.

BPC Roles

The departments of State and Defense, as well as USAID, all play important roles in the development and execution of U.S. efforts to build partner capacity. This section provides a brief overview on BPC roles and guidance, and is not intended to be a comprehensive primer on all BPC initiatives.

Department of State

The Department of State leads U.S. interagency policy initiatives and oversees policy and programmatic support for security cooperation and security assistance programs through its bureaus, offices, and overseas missions as directed by the President in NSPD-1, and it leads integrated U.S. government reconstruction and stabilization efforts as directed by NSPD-44 (which will be discussed in greater detail later in this chapter). The Department of State's responsibilities also include oversight of other U.S. government foreign policies and programs that may have an impact on the security sector.

In particular, State's Bureau of Political-Military Affairs is responsible for providing policy direction in the areas of international security, security assistance, military operations, defense strategy and plans, and defense trade. Among its many responsibilities, the Bureau of Political-Military Affairs plays a key role in achieving peace and security around the world by:[1]

- Countering terrorism and responding to crises by managing and sustaining coalitions, working with DoD on strategic and contingency planning to include counterinsurgency policy, and reinforcing the capabilities of friends and allies to respond to respond to humanitarian and natural disasters.
- Managing and regulating defense trade and arms transfers to reinforce the military capabilities of friends, allies, and coalition partners, and to ensure that the transfer of U.S.-origin defense equipment and technology supports U.S. national security interests.
- Promoting regional security through bilateral and multilateral cooperation and dialogue, as well as through the provision of security assistance to friendly countries and international peacekeeping efforts.
- Providing diplomatic support to U.S. military operations, including the negotiation of status of forces, defense cooperation, base access, cost-sharing, and nonsurrender agreements.
- Countering the destructive effects of conventional weapons by clearing landmines and reducing the availability of at-risk small arms and light weapons, including man-portable air defense systems.

U.S. Agency for International Development

The U.S. Agency for International Development carries out a variety of economic assistance programs designed to help the people of certain less-developed countries develop their human and economic resources,

---

[1] Department of State, Bureau of Political-Military Affairs web site.

increase productive capacities, and improve the quality of human life as well as to promote economic and political stability in friendly countries. USAID performs its functions under the direction and foreign policy guidance of the Secretary of State. The agency is charged with central direction and responsibility for the U.S. foreign economic assistance program. In relation to BPC, USAID's primary role is to support governance, conflict mitigation and response, reintegration and reconciliation, and rule-of-law programs aimed at building civilian capacity to manage, oversee, and provide security and justice.

Department of Defense

In contrast to DOS and USAID, in recent years DoD has developed formal guidance on its role in BPC missions. In particular, the Office of the Secretary of Defense (OSD) BPC Execution Roadmap from the 2006 Quadrennial Defense Review emphasizes building the military capabilities of partner countries that will enable them to make valuable contributions to coalition operations.[2] The BPC Execution Roadmap not only includes guidance on how DoD should train and equip foreign military forces, but also discusses the need to improve the capacity of other nonmilitary security services (i.e., stability police, border guards, customs, etc.) within partner countries. Moreover, the concept also refers to the need to improve DoD's ability to work with other U.S. government departments and agencies, nongovernmental organizations (NGOs), and the private sector in an operational context for integrated operations.

The BPC Execution Roadmap directly connects stability operations to DoD security cooperation, though not in great detail, as it directs DoD stakeholders conducting security cooperation activities to

- Improve partner capabilities to enhance their prospects for stability mission success.
- Improve partner capabilities to reduce stability burdens on U.S. forces.

---

[2]   See Moroney et al., *Building Partner Capabilities for Coalition Operations*, 2007.

The BPC Execution Roadmap leaves much room for interpretation, as it does not specify which capabilities should be focused on, which allies and partners to work with, or which security cooperation tools are best suited to build such capabilities with the respective partners. In addition, the BPC Execution Roadmap does not provide specific guidance on military capabilities to be cultivated in partners for stability operations, nor does it specify key allies and partner countries to work with to achieve common ends. That guidance is found elsewhere, as discussed below.

Two additional sources of guidance that shape the way DoD conducts BPC activities in general are the OSD Guidance for the Employment of the Force, which COCOMs use as the basis for developing combatant command Theater Security Cooperation Plans.[3] These documents do identify specific countries to work with and priorities for the types of capacity to build.

Stability Operations Roles and Missions

Stability operations necessarily draw on all elements of national power: diplomatic, military, information, and economic. The National Security Strategy and National Security Presidential Directive 44 take this into account when assigning roles and responsibilities for the conduct of stability operations, and NSPD-44, in particular, directs U.S. agencies to work together to plan and execute post-conflict reconstruction missions.[4]

Table 2.1 shows the main focus of State Department and USAID versus the focus of DoD across the three main sectors of conflict.

---

[3]   The OSD Guidance for the Employment of the Force replaces the Security Cooperation Guidance and merges with the Contingency Planning Guidance.

[4]   NSPD-44.

Table 2.1
Organizational Roles Across the Main Phases of Conflict

	Conflict Prevention	Conflict Management	Post-Conflict Reconstruction
Lead organization	State Department/ USAID	Department of Defense	State Department
Supporting organization	Department of Defense	State Department	Department of Defense

The following sections examine the roles and missions assigned to the State Department, USAID, and DoD in the planning and execution of stability operations.

Department of State

The State Department, specifically the Coordinator for Reconstruction and Stability, is designated by NSPD-44 as the focal point for coordinating reconstruction and stabilization efforts. According to NSPD-44, DOS is responsible for developing "strategies to build partnership security capacity abroad and seek to maximize nongovernmental and international resources for reconstruction and stabilization activities."[5] According to S/CRS officials, State's goals for stability operations include managing underlying tensions while laying the groundwork for long-term development; rebuilding the political, socioeconomic, and physical infrastructure of a country; working to diminish the drivers of conflict while developing local capacity to govern; and helping to transfer power back to an indigenous government.

Currently, the State Department focuses its stability operations mission on post-conflict reconstruction and stabilization or nation building after external intervention. To do this, State identifies these five broad sectors that describe stability missions:

- Security.
- Governance and participation.
- Justice and reconciliation.

---

[5]   NSPD-44, p. 4.

- Humanitarian assistance and social well-being.
- Economic stabilization and infrastructure.

The State Department's primary source for stability-related tasks and required capabilities is the Essential Tasks Matrix, which is organized into the five broad technical sectors listed above.[6] Because NSPD-44 has designated State as the lead agency to coordinate and integrate U.S. government agencies to prepare, plan, and conduct stability operations activities,[7] the ETM list is increasingly emerging as the accepted list of stability operation tasks.[8] These technical sectors have been adopted directly by emerging joint peace operations doctrine and are broadly reflected by the DoD's six major mission elements for a stability operation (which will be detailed in the DoD subsection below).

However, the State Department is understaffed and underfunded to direct large stability operations. In essence, the military has the personnel and staffing to direct BPC for stability operations, while State has the authority and oversight mandate. This has led to many problems in the interagency coordination of this essential task.

U.S. Agency for International Development

Unlike the State Department, USAID is not tasked specifically by NSPD-44 to implement its provisions. Rather, USAID is one of the "other agencies and departments" referred to in the document, all of which should coordinate with DOS. In doing so, USAID's goal in stability operations is to reverse the decline in fragile states and advance their recovery to a stage where transformational development progress is possible. USAID officials recognize that a unilateral approach will not be sufficient to address the complex challenges of fragile states and

---

[6] S/CRS, *Post-Conflict Reconstruction Essential Tasks,* April 2005.

[7] NSPD-44, p. 2.

[8] JP 3-07.3, *Peace Operations,* equates peace building with stability operations and adopts the ETM's five basic mission areas. It also directs the reader to the ETM "for detailed descriptions of tasks and considerations" within each mission sector. JCS, *Peace Operations,* JP 3-07.3, October 2007, pp. IV-1 to IV-2, IV-14. The Army's recently published FM 3-07, *Stability Operations,* draws heavily upon the ETM for its list of essential stability tasks.

a coordinated U.S. government approach will be necessary. As a result, in 2005 USAID created the Office of Military Affairs to serve as the USAID-specific entity to support an integrated interagency approach. This change, along with USAID's efforts to coordinate with the departments of State, Defense, Treasury, Justice, and others, should help to ensure that diplomatic, security, and military efforts are mutually reinforced and continue to improve.

Department of Defense

The Department of Defense plays a leading role in the implementation of stability operations strategy. In particular, DoD supports the implementation of NSPD-44 though DoD Directive (DoDD) 3000.05.[9] DoDD 3000.05 essentially does three things. First, it identifies goals and tasks as listed below.

Stability Operations Goals:

- Provide local populace with security.
- Restore essential services.
- Meet humanitarian needs.
- Develop indigenous capacity for restoring essential services.
- Develop indigenous capacity for viable market economy.
- Develop indigenous capacity for rule of law.
- Develop indigenous capacity for democratic institutions.
- Develop indigenous capacity for civil society.

Stability Operations Tasks:

- Rebuild indigenous institutions including various types of security forces, correctional facilities, and judicial systems necessary to secure and stabilize the environment.
- Revive or build the private sector, by encouraging citizen-driven, bottom-up economic activity and constructing necessary infrastructure.

---

[9]   DoDD 3000.05, *Military Support for SSTR Operations*. This directive states stability operations shall be given the same priority as combat operations and that they will be explicitly addressed in all DoD doctrine, organizations, training, education, exercises, materiel, leadership, personnel, facilities, and planning activities.

- Develop representative governmental institutions.

Second, DoDD 3000.05 assigns responsibility for implementing stability operations within DoD and directs the services and components to coordinate and train with interagency and multinational partners as well as NGOs. It also assigns the military departments the responsibility to develop stability operations capabilities. Third, it tasks the Under Secretary of Defense for Policy to develop a list of countries and areas with the potential for U.S. military engagement in stability operations, which is defined as "military and civilian activities conducted across the spectrum from peace to conflict to establish or maintain order in states and regions."[10]

In particular, DoDD 3000.05 directs the military to organize in order to accomplish two broad missions:

- Build sustainable peace.
- Advance U.S. interests.

To help plan for these missions, Joint Forces Command developed a joint operating concept for military support to stabilization, security, transition, and reconstruction operations, which describes a whole range of operational objectives that span the spectrum of conflict activities. This JOC describes how future joint force commanders will provide military support to stability operations within a military campaign. Additionally, this JOC identifies the operational capabilities required for achieving military campaign objectives and effects in support of national strategic end states.

The stability operations JOC clearly delineates the difference between military support to civilian authorities in the conflict prevention and post-conflict reconstruction phases, and adds a third mission—that of conflict management. This includes military-led operations such as major combat operations, as well as nontraditional U.S. military missions such as civil unrest, insurgency, terrorism, and factional conflict. Because this conflict management mission is not

---

[10] DoDD 3000.05, *Military Support for SSTR Operations*.

addressed in DoDD 3000.05, we do not include it in the discussion on BPC for stability operations.[11]

The JOC for military support to SSTR operations identifies six major mission elements, which are:

- **Establish and maintain a safe and secure environment.** The purpose of this MME is to create a situation in which the security of the people, property, and livelihoods within the country is sufficient to allow the general populace to routinely go about its business.[12]
- **Establish representative, effective governance and the rule of law.** The purpose of this MME is to establish and maintain the institutions and processes required for representative and effective local and national governance that is accepted as legitimate by the indigenous population.[13]
- **Deliver humanitarian assistance.** This MME involves the provision of immediate and emergency life support to populations where serious threats to life and property exist.
- **Reconstruct critical infrastructure and restore essential services.** This MME focuses on less-immediate life support tasks that meet the basic needs of the indigenous population and which are intended to prevent loss of life and the spread of instability or insurgency.
- **Support economic development.** This MME focuses on more traditional developmental tasks that are intended to reduce potential long-term drivers of instability within a country.
- **Conduct strategic communications.** This MME is intended to understand and engage key local and foreign audiences in order to create, strengthen, or preserve conditions favorable to achievement of overall stability operations goals and objectives.[14]

---

[11] DoD, *Military Support to SSTR Operations JOC*, Version 2.0, December 2006.

[12] DoD, *Military Support to SSTR Operations JOC*, Version 2.0, December 2006, p. 33.

[13] Derived from DoD, *Military Support to SSTR Operations JOC*, Version 2.0, December 2006, p. 61.

[14] DoD, *Military Support to SSTR Operations JOC*, Version 2.0, December 2006, p. vi.

There are differences in the stability operations lexicon used by the DoD and DOS. Joint Publication 3-07.3, *Peace Operations,* uses the same terminology as the State Department's ETM, but specifically places the re-establishment of critical infrastructure and essential services within the economic stabilization and infrastructure MME.[15] Therefore, we have adopted a modified version of the stability operations JOC taxonomy for methodological and analytical reasons because it is better suited for determining required capabilities than State's ETM taxonomy.[16] Table 2.2 provides our alignment of the State Department's ETM technical sector list with DoD's list of MMEs.

Table 2.2
Comparison of State Department ETM Technical Sectors and DoD's Stability Operations MMEs

ETM Technical Sector (DOS)	Major Mission Element (DoD)
Security	Establish and maintain a safe and secure environment
Governance and participation	Conduct strategic communications
Justice and reconciliation	Establish representative, effective governance and the rule of law
Humanitarian assistance and social well-being	Deliver humanitarian assistance
	Reconstruct critical infrastructure and restore essential services
Economic stabilization and infrastructure	Support economic development

SOURCE: DoD, *Military Support to SSTR Operations JOC,* December 2006, pp. iv, 21; DOS, *Post-Conflict Reconstruction Essential Tasks,* April 2005.

NOTE: Although the "conduct strategic communications" MME is aligned with "governance and participation" from the DOS ETM, strategic communications infuses all of the ETM sectors.

---

[15] JCS, *Peace Operations,* JP 3-07.3, October 2007, p. IV-9.

[16] These temporal categories are immediate response, transition, and fostering sustainability.

The Intersection of BPC and Stability Operations Guidance for the U.S. Army

The U.S. Army is the service most likely to be tasked to conduct stability operations. Responding to these various DOS and DoD guidance documents presents a challenge for the U.S. Army. Although there is a lack of guidance specifically addressing BPC for stability operations at either the DOS or DoD levels, the U.S. Army integrated guidance from the higher levels to develop its own security cooperation strategy and BPC guidance for stability operations. In particular, the Army has published this guidance in Army FM 3-0, *Operations,* and in FM 3-07, *Stability Operations,* which will be described below. The Army also translates OSD and COCOM guidance on security cooperation and BPC into its Army Security Cooperation Strategy, which provides guidance to the Army service component commands and other Army major commands on Army priorities for security cooperation. Figure 2.1 depicts the relationships between the various BPC and stability

Figure 2.1
The Army's Integration of BPC and Stability Operations Guidance

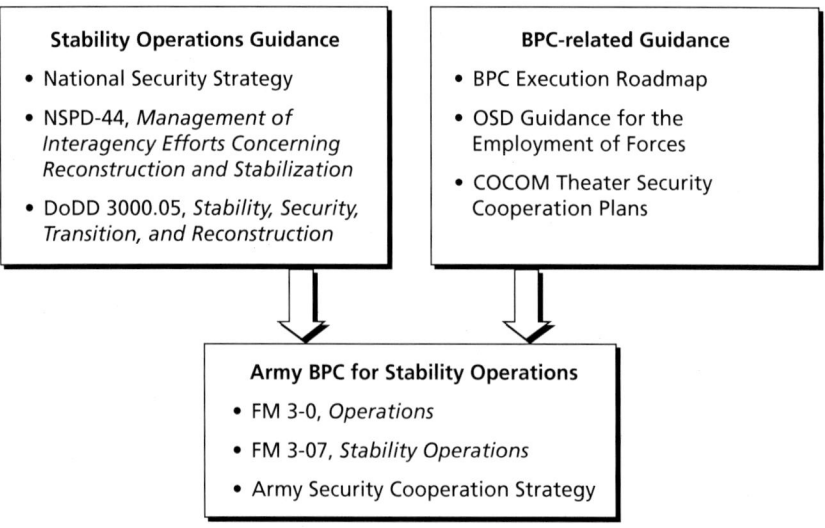

operations guidance documents, focusing on how the Army draws on them to develop its own guidance for building partner stability operations capacity.

Army Field Manuals 3-0 and 3-07

In response to operational requirements and DoDD 3000.05 guidance, the Army has undertaken significant doctrinal efforts to outline specific stability operations missions and tasks. FM 3-0, *Operations*, is the Army's capstone doctrinal document that provides the "overarching doctrinal guidance and direction for conducting operations." It provides the Army's view of how operations should be conducted and lays the foundation for all other doctrinal developments as well as providing a guide for shaping the future development of the Army.[17] FM 3-0 formally establishes stability operations as being as important as offensive and defensive operations and describes the primary military task to be conducted to support broader U.S. government stability efforts.[18] The stability tasks outlined in FM 3-0 are fleshed out in FM 3-07, *Stability Operations*, which provides the overarching doctrinal guidance and direction for the Army's conduct of stability operations.[19] FM 3-07 addresses the role of Army stability operations in the context of broader U.S. government objectives and describes the Army's role in using its capabilities to establish a safe and secure environment that will enable a return to civil authority.[20]

Army Security Cooperation Strategy

Although the Army Security Cooperation Strategy provides guidance on Army priorities for security cooperation, it does not identify stability operations partners, as such. Rather, it identifies priority countries for the Army to work with in more general security cooperation terms, based on OSD and COCOM guidance, and on Army priorities.

---

[17] HQDA, *Operations*, FM 3-0, February 2008, pp. v, vii; General Casey, *Opening Remarks, Senate Armed Service Committee*, 26 February 2008.

[18] HQDA, *Operations*, FM 3-0, February 2008, pp. vii, viii, 3-1, 3-12 to 3-17.

[19] HQDA, *Stability Operations*, FM 3-07, October 2008, p. iv.

[20] HQDA, *Stability Operations*, FM 3-07, October 2008, pp. vi to vii.

Despite the lack of specific guidance that explicitly connects stability operations with BPC, Army-led workshops on stability operations almost always include some aspect of the need to build the capacity of partner armies for stability (or, more specifically, for "stability operations," which is the Army's term of preference).[21]

Relevant U.S. Army Capabilities for Stability Operations

Currently the Army's stated purpose for stability operations is to create security conditions that "allow the other instruments of state power to become preeminent."[22] This objective is accomplished by a combination of five primary stability tasks, also called the Army's "logical lines of operation" (LLOs):

- **Civil security.** Focuses on the protection of the populace from internal and external threats.
- **Civil control.** Focuses on regulating the behavior and activities of groups in order to allow for the provision of security and essential services while military forces are conducting operations.
- **Restoration of essential services.** Focuses on establishing or restoring basic services and protecting them until local or host-nation authorities can provide them.
- **Support to governance.** Focuses on establishing conditions that allow for the transfer of authority to civilian and host-nation agencies.
- **Support to economic and infrastructure development.** Focuses on supporting economic and infrastructure development that helps develop host-nation capability and capacity in these areas.[23]

---

[21] Examples of such workshops include those hosted by the Army Peacekeeping and Stability Operations Institute at the Army War College and the Army UNIFIED QUEST Title 10 exercise series. RAND Arroyo Center study team members attended these events.

[22] HQDA, *Operations*, FM 3-0, p. 3-14.

[23] HQDA, *Operations*, FM 3-0, pp. 3-12 to 3-14.

These LLOs align with the stability operations JOC end state of reaching "full host nation responsibility across the MMEs in the context of a new domestic order resolving earlier sources of instability to ensure a viable, sustainable peace," and they square the lack of DoD guidance on stability operations missions with the heavy focus from the State Department on post-conflict reconstruction.[24] In addition, the five primary stability tasks are roughly equivalent to the State Department's ETM sectors and DoD's MMEs.[25]

Given the State Department ETMs and DoD MMEs for stability operations discussed above, the study team was able to align the Army's LLOs, as shown in Table 2.3. As a result, this mapping enabled the team to identify the necessary stability operations capabilities that are directly relevant to the broad Army missions.

Table 2.3
Mapping the State Department ETM Sectors, DoD Major Mission Elements, and Army Logical Lines of Operation

ETM Technical Sector (DOS)	Major Mission Element (DoD)	Logical Lines of Operation (Army)
Security	Establish and maintain a safe and secure environment	Civil security
Governance and participation	Conduct strategic communications	Support to governance
	Establish representative, effective governance and the rule of law	
Justice and reconciliation		Civil control
Humanitarian assistance and social well-being	Deliver humanitarian assistance	Restore essential services
	Reconstruct critical infrastructure and restore essential services	
Economic stabilization and infrastructure	Support economic development	Support economic and infrastructure development
END STATE: The establishment of a new domestic order resolving earlier sources of instability to ensure a viable and sustainable peace.		

SOURCES: Derived from DoD, *Military Support to SSTR Operations JOC,* December 2006, pp. iv, 21; S/CRS, *Post-Conflict Reconstruction Essential Tasks,* April 2005.

---

[24] DoD, *Military Support to SSTR Operations JOC,* December 2006, p. C-1.

[25] HQDA, *Operations,* FM 3-0, February 2008, Figure 3-3.

The ETM sectors and MMEs help organize the types of capabilities that the Army needs to develop in order to be able to conduct a wide range of activities in support of broad stability goals. In many cases, the types of capabilities needed for these missions will be similar. For example, providing shelter for refugees will be similar when conducting either a disaster relief mission or a post-conflict reconstruction mission. On the other hand, it is important to be able to discern between tasks conducted during peacetime and the same tasks conducted during irregular warfare or major combat operations. Providing for refugees in the middle of a war will require a higher level of support from military forces than civilians. Therefore, it is not enough to focus on building a capability without also thinking about how those capabilities will be utilized in support of which missions.

The following sections introduce and define the critical capabilities associated with the stability operations MMEs list above. Each MME section includes working definitions of required stability operations capabilities, which are drawn from official DOS and DoD guidance. Appendix A expands upon this information by providing more detailed definitions for these required capabilities, lists the State Department sectoral subtasks associated with those capabilities, and, where necessary, provides more information as to how we constructed the MMEs.

Establish and Maintain a Safe and Secure Environment

Of the six MMEs, establishing and maintaining a safe and secure environment is the most relevant one for the Army, and the most likely to call on Army capabilities for its implementation. The primary purpose of this MME is to create a situation where the security within a country is sufficient to allow the general populace to routinely go about its business. A safe, secure environment facilitates the conduct of large-scale, civilian-led external assistance efforts, as well as host-nation activities for reconstruction.[26] A key part of this process is the development a self-sustaining public law-and-order system operating in accordance with internationally recognized standards and with respect for

---

[26] DoD, *Military Support to SSTR Operations JOC*, December 2006, p. 33.

internationally recognized human rights and freedoms. Civilian organizations are primarily responsible for civil law and order and have the primary responsibility to work with the host nation to train, advise, and support their efforts to establish a viable rule-of-law system and facilitate social recovery.[27]

Several capabilities flow from this MME:

- **Conduct peace operations.** Conduct tactical military operations designed to monitor, facilitate, or enforce the implementation of an agreement, either negotiated or imposed, that are intended to create the condition for conflict resolution in order to establish and maintain peace. This includes ceasefires, truces, or other such agreements.[28]
- **Conduct disarmament, demobilization, and reintegration operations** in support of war-to-peace transitions by reducing or eliminating belligerent armed forces and the supply of armed weapons and through facilitating the return of ex-combatants to sustainable civilian livelihoods.
- **Develop and sustain armed services and intelligence forces** that can conduct legitimate self-defense operations to maintain control or regain control over national territory. This includes the ability to create professional military and intelligence forces that are transparent and accountable to the civilian government.[29]
- **Establish and maintain border and boundary control** and regulate the movement of people and goods across them.[30]

---

[27] Derived from JCS, *Peace Operations,* JP 3-07.3, October 2007.

[28] Task OP 3.3.1, Conduct Peacekeeping Operations in the Joint Operations Area; task OP 3.3.2, Conduct Peace Enforcement Operations in the Joint Operations Area. From JCS, *Universal Joint Task List,* CJCSM 3500.04D, August 2005, p. B-C-C-69.

[29] Peace and Security program subelement 3.6.1, Territorial Security and Governing Justly and Democratically; program element 2.5, Governance of the Security Sector. From *Foreign Assistance Standardized Program Structure and Definitions,* 20 October 2006.

[30] Derived from the FM 3-07.31 definition of border control. ALSA Center, *Multi-Service Tactics, Techniques and Procedures for Conducting Peace Operations,* FM 3-07.31, October 2003, p. III-1.

- **Establish and maintain freedom of movement.** Ensure the uninhibited movement of civilian traffic and commerce so as to allow the resumption of normal activity and to guarantee the right of transit of NGOs, noncombatants, and stability operations personnel.[31]
- **Establish an identification regime.** Plan, establish, and enforce a civilian identification regime, including documents relating to personal identification, property ownership, court records, voter registries, birth certificates, and driving licenses.[32]
- **Provide interim public order.** Ensure a lawful and orderly environment and suppress criminal behavior. This includes the ability to protect vulnerable noncombatants and to engage in crowd and disturbance control operations.[33]
- **Conduct civilian police operations.** Establish and sustain effective, professional, and accountable law enforcement services with the capacity to protect persons, property, and democratic institutions against criminal and other extralegal elements.[34]
- **Conduct emergency clearance operations** to remove or neutralize mines and unexploded ordnance that are an immediate threat to civilians and stability operations personnel.[35]
- **Provide protective services.** Protect key political and societal leaders from assassination, kidnapping, injury, or embarrassment.[36]

---

[31] HQDA, *Stability Operations and Support Operations*, FM 3-07, February 2003, pp. 2-2, 4-9 to 4-10, 4-17.

[32] This is a part of population and resource control operations. See HQDA, *Counterinsurgency*, FM 3-24, December 2006, p. 5-21; HQDA, *Army Universal Task List*, FM 7-15, July 2006, pp. 6-112 to 6-113.

[33] Derived from the definition for Army Tactical Task 7.7.2.2, Provide Law and Order. HQDA, *Army Universal Task List*, FM 7-15, July 2006, p. 7-41.

[34] Derived from the Peace and Security program subelement 3.7, Law Enforcement Reform, Restructuring, and Operations definition in *Foreign Assistance Standardized Program Structure and Definitions*, 20 October 2006.

[35] HQDA, *Army Universal Task List*, FM 7-15, July 2006, p. 5-4; HQDA, *Stability Operations and Support Operations*, FM 3-07, February 2003, pp. 2-2, 2-8 to 2-9.

[36] Derived from the definition for Army Tactical Task 5.3.6.1, Provide Protective Services for Selected Individuals. HQDA, *Army Universal Task List*, FM 7-15, July 2006, p. 5-74.

- **Protect critical installations and facilities** from hostile actions. This includes securing and protecting private property and factories, religious sites, cultural sites, military facilities, critical infrastructure and natural resources, and public institutions.
- **Protect reconstruction and stabilization personnel.** Provide physical security and logistical support for civilian personnel and facilities engaged in stability operations.[37]
- **Coordinate indigenous and international security forces and intelligence support** for the purposes of accomplishing the operations objectives.[38] This includes the ability to integrate command, control, and intelligence and information sharing arrangements between international military, constabulary, and civilian police forces and between the international and indigenous security forces.
- **Participate in stability operations-related regional security arrangements.** Negotiate, participate in, and comply with regional security arrangements, including those that enhance border security and control as well as regional security.

Establish Representative, Effective Governance and the Rule of Law

The objective of this MME is to establish and maintain the institutions and processes required for representative and effective local and national governance that the indigenous population accepts as legitimate.[39] The development of effective governing institutions is a key

---

[37] Derived from Peace and Security program subelement 3.1.6, Armed Physical Security, in *Foreign Assistance Standardized Program Structure and Definitions*, 20 October 2006. See also task ST 4.3.2, Provide Supplies and Services for Theater Forces, JCS, *Universal Joint Task List*, CJCSM 3500.04D, August 2005, p. B-C-B-66 and HQDA, *Stability Operations and Support Operations*, FM 3-07, February 2003, p. 4-10.

[38] Derived from the Joint Interagency/international/multinational/NGO Coordination tier 1 joint capability area. *Joint Capability Areas Tier 1 and Supporting Tier 2 Lexicon: Post 24 August 2006 JROC*, August 2006, pp. 42–43.

[39] Derived from the DoD definition for establishing representative, effective government and the rule of law. DoD, *Military Support to SSTR Operations JOC*, Version 2.0, December 2006, p. 61.

requirement for establishing government legitimacy and is important for establishing lasting stability.[40] This includes the existence of meaningful avenues of public participation and oversight, substantive separation of powers through institutional checks and balances, and governmental transparency and integrity, which is a key component of government effectiveness and political stability.[41] These capabilities are primarily civilian-led tasks.

- **Establish a temporary civil administration** until an effective indigenous or local government can be constituted.[42]
- **Establish executive authority.** Establish, develop, and maintain executive offices, ministries, and independent governmental bodies that operate efficiently and effectively, incorporate democratic principles, are responsive to the public, are accountable, and which can implement and enforce laws, regulations, and policies.[43]
- **Establish, develop, and maintain legislatures and legislative processes** that uphold democratic practices, produce effective legislation and regulations, are responsive to the populace, encourage public participation in policymaking, hold themselves and the executive branch accountable, and oversee the implementation of government programs, budgets, and laws.[44]

---

[40] HQDA, FM 3-24, December 2006, p. 5-15.

[41] Derived from the Governing Justly and Democratically program area 2, Good Governance, definition in *Foreign Assistance Standardized Program Structure and Definitions*, 20 October 2006.

[42] Derived from Army Tactical Task 6.16.6, Establish Temporary Civil Administration. HQDA, *Army Universal Task List*, FM 7-15, July 2006, p. 6-120; HQDA, *Civil Affairs Operations*, FM 41-10, February 2000, pp. 2-27 to 2-33, G-3.

[43] Derived from the Governing Justly and Democratically program element 2.2, Public Sector Executive Function, definition in *Foreign Assistance Standardized Program Structure and Definitions*, 20 October 2006.

[44] Derived from the Governing Justly and Democratically program element 2.1, Legislative Function and Processes, definition in *Foreign Assistance Standardized Program Structure and Definitions*, 20 October 2006.

- **Assist local governance** to effectively plan, manage, finance, deliver, and account for local public goods and services.
- **Enhance transparency and anti-corruption.** Make transparent and accountable the government institutions, processes, and policies. This includes the capability to enforce anti-corruption laws and regulations.[45]
- **Conduct legitimate elections** that are a legitimate contestation of ideas and political power and which reflect the will of the people. This includes the capability to establish, develop, and maintain a legal and regulatory framework that allows political parties and entities to operate within a competitive multiparty system.[46]
- **Help establish, develop, and sustain viable political parties** and political entities that are effective and accountable, that represent and respond to citizens' interests, and that govern responsibly and effectively.[47]
- **Build civil society.** Enable citizens to freely organize, advocate, and communicate with their government and with each other.[48]
- **Build a free media.** Establish, develop, and sustain a broadly functioning independent media sector that can reinforce and foster democratic governance.[49]

---

[45] Derived from the Governing Justly and Democratically program area 4, Civil Society, definition in *Foreign Assistance Standardized Program Structure and Definitions*, 20 October 2006.

[46] Derived from the Governing Justly and Democratically program element 3.2, Elections and Political Processes, definition in *Foreign Assistance Standardized Program Structure and Definitions*, 20 October 2006.

[47] Derived from the Governing Justly and Democratically program element 3.3, Democratic Political Parties, definition in *Foreign Assistance Standardized Program Structure and Definitions*, 20 October 2006.

[48] Derived from the Governing Justly and Democratically program element 2.4, Anti-Corruption Reforms, definition in *Foreign Assistance Standardized Program Structure and Definitions*, 20 October 2006.

[49] Derived from the Governing Justly and Democratically program element 4.2, Media Freedom and Freedom of Information, definition in *Foreign Assistance Standardized Program Structure and Definitions*, 20 October 2006.

- **Provide an interim criminal justice system** capable of sustaining law and order until an indigenous capacity to do so has been developed or restored.[50]
- **Provide judicial personnel and infrastructure.** Establish, develop, and maintain an effective, accountable, and procedurally fair civil and criminal justice institution as well as provide the personnel required for its operation. The system should be capable of ensuring equality before the law by conducting fair trials.
- **Establish, maintain, and operate a fair, transparent, and accountable corrections system** that complies with international human rights standards.[51]
- **Foster legal system reform.** Develop and sustain a democratic legal and regulatory framework that is consistent with international human rights standards.[52]
- **Enforce property rights.** Establish or improve transparent, equitable, and accountable institutions that resolve property disputes and enforce property rights.[53]
- **Safeguard human rights.** Protect, promote, and enforce internationally recognized human rights standards.[54]
- **Conduct programs to combat human trafficking.** Develop, execute, and sustain anti-trafficking programs and to provide support for and the protection of trafficking victims.[55]

---

[50] Derived from JP 3-07.3, *Peace Operations,* October 2007.

[51] Derived from the Governing Justly and Democratically program element 1.3, Justice System, and Peace and Security program subelement 3.1.2, Corrections Assistance, definitions in *Foreign Assistance Standardized Program Structure and Definitions,* 20 October 2006.

[52] Derived from the Governing Justly and Democratically program element 1.1, Constitutions, Laws and Legal Systems, definition in *Foreign Assistance Standardized Program Structure and Definitions,* 20 October 2006.

[53] Derived from the Economic Growth program subelement 6.1.1, Property Rights, definition in *Foreign Assistance Standardized Program Structure and Definitions,* 20 October 2006.

[54] Derived from Governing Justly and Democratically program subelement 1.4, Human Rights, in *Foreign Assistance Standardized Program Structure and Definitions,* 20 October 2006.

[55] Derived from the Peace and Security program element 5.3, Trafficking-In-Persons and Migrant Smuggling, definition in *Foreign Assistance Standardized Program Structure and Definitions,* 20 October 2006.

- **Support reconciliation** to address past human rights abuses and social traumas through legal procedures that build respect for the rule of law. This is also intended to promote justice, psychological relief and reconciliation in order to achieve a sustainable peace.[56]
- **Address past war crimes and human rights violations** through retributive justice mechanisms such as war crimes courts and tribunals that are transparent, accountable, and conform to international legal norms.[57]
- **Establish truth commissions and support remembrance.** Address past war crimes and human rights violations through restorative justice mechanisms such as truth and reconciliation commissions and reparations.[58]
- **Community rebuilding.** The ability to provide the local populace with the means to form a cohesive society.[59]

Deliver Humanitarian Assistance

The objective of this MME is to rapidly relieve or reduce the results of natural or man-made disasters or other endemic conditions such as human suffering, disease, or privation that might represent a serious threat to life or that can result in great damage to or loss of property through the delivery of humanitarian assistance.[60] Such operations are intended to be emergency in nature, and while they should help create the foundations for long-term recovery and development, they are not a substitute for the development investments required to reduce chronic

---

[56] Derived from JP 3-07.3, *Peace Operations*, October 2007, p. IV-8.

[57] Derived from the Governing Justly and Democratically program subelement 1.1.3, Transitional Justice, definition in *Foreign Assistance Standardized Program Structure and Definitions*, 20 October 2006.

[58] Derived from the Governing Justly and Democratically program subelement 1.1.3, Transitional Justice, definition in *Foreign Assistance Standardized Program Structure and Definitions*, 20 October 2006.

[59] HQDA, *Stability Operations*, FM 3-07, October 2008, p. 3-9.

[60] JCS, *Peace Operations,* JP 3-07.3, October 2007, p. IV-5; DoD, *Military Support to SSTR Operations JOC*, Version 2.0, December 2006, p. 42.

poverty or establish social services.[61] The effective delivery of humanitarian assistance requires the ability to obtain and redistribute essential supplies, food, and medicine within an affected region, or deliver essential items that are not available locally or regionally to the disaster sites.[62]

The required capabilities for this MME are listed below and are drawn primarily from the stability operations JOC, but also include elements from the State Department's ETM.

- **Conduct refugee and internally displaced persons operations.** Plan, construct, and operate camps and facilities for refugees and internally displaced persons.[63]
- **Provide emergency power supply.** Promptly deliver, operate, and maintain electrical power generation equipment to affected regions.[64]
- **Provide emergency water supply and sanitation services.** Promptly deliver, operate, and maintain emergency water purification, water distribution systems, and meet basic sanitation standards in the affected regions.[65]
- **Provide emergency food and non-food relief.** Promptly deliver and distribute emergency food and non-food supplies to affected regions.[66]

---

[61] Derived from the Humanitarian Assistance definition in *Foreign Assistance Standardized Program Structure and Definitions*, 20 October 2006.

[62] DoD, *Military Support to SSTR Operations JOC*, Version 2.0, December 2006, p. 59.

[63] Derived from DoD, *Military Support to SSTR Operations JOC*, Version 2.0, December 2006, p. 60; and DOS, *Post-Conflict Reconstruction Essential Tasks*.

[64] Derived from DoD, *Military Support to SSTR Operations JOC*, Version 2.0, December 2006, p. 59.

[65] Derived from DoD, *Military Support to SSTR Operations JOC*, Version 2.0, December 2006, p. 59; and DOS, Humanitarian Assistance program subelement 1.2.2, Water and Sanitation Commodities and Services, definition in *Foreign Assistance Standardized Program Structure and Definitions*, 20 October 2006.

[66] Derived from the S/CRS ETM, shelter construction.

- **Provide emergency shelter.** Plan and execute emergency shelter programs and deliver the required supplies in the affected regions.[67]
- **Provide emergency medical treatment.** Provide timely emergency medical treatment and prophylaxis to people affected by natural or man-made disasters.[68]
- **Conduct humanitarian de-mining operations.** Completely remove all mines and unexploded ordnance after the end of hostilities in order to safeguard the civilian population within a geopolitical boundary.[69]

Reconstruct Critical Infrastructure and Restore Essential Services

The objective of this MME is to address the life support need of the indigenous population. In an unstable environment, the U.S. military may initially have the lead role in this task, as other agencies may not be present or may lack the capability and capacity to meet the needs of the indigenous population. Due to uncertainties in the security environment, the military must be prepared to perform these tasks for an extended period and under difficult security circumstances[70] in an effort to prevent the loss of life and the spread of insurgency.[71]

- **Restore, establish, and maintain firefighting services** capable of a timely response to property fires.[72]

---

[67] Derived from the S/CRS ETM, shelter construction.

[68] DoD, *Military Support to SSTR Operations JOC*, Version 2.0, December 2006, p. 60.

[69] Derived from the definition for de-mining in FM 20-32. HQDA, *Mine/Countermine Operations*, FM 20-32, October 2002, pp. 9-2, 9-7. See also Peace and Security program subelement 3.4, Explosive Remnants of War, in *Foreign Assistance Standardized Program Structure and Definitions*, 20 October 2006. JP 3-07.3 categorizes humanitarian de-mining as a security function. JCS, *Peace Operations,* JP 3-07.3, October 2007, p. IV-4.

[70] HQDA, FM 3-24, 2006, pp. 5-14 to 5-15.

[71] JCS, *Peace Operations,* JP 3-07.3, October 2007, p. IV-9.

[72] Derived from HQDA, FM 3-24, 2006, p. 5-15.

- **Build, restore, maintain, and operate water purification plants and potable water distribution systems.**[73] The primary objective of this ability is to ensure that water treatment plants and the distribution systems for potable water are functional.[74]
- **Build, restore, maintain, and operate power generation grids** to ensure the local distribution of electrical power.[75]
- **Build, restore, maintain, and operate schools and universities.**[76] The primary objective of this capability is to ensure that schools and universities are open, staffed, and supplied.[77]
- **Repair and maintain transportation networks.** Repair, construct, maintain, and operate roads, bridges, tunnels, ports, and airfields for road, rail, air, and sea transportation.[78]
- **Repair and maintain public health facilities.** Repair, build, maintain, and operate primary health care clinics, hospitals, and other elements of the health care system.[79]
- **Maintain public sanitation.** In the uncertain aftermath of a natural disaster, man-made disaster, or conflict, the ability to repair, construct, maintain, and operate sewage disposal systems and collect and dispose of garbage.[80]

---

[73] Derived from DoD, *Military Support to SSTR Operations JOC*, Version 2.0, December 2006, p. 60.

[74] HQDA, FM 3-24, 2006, p. 5-15.

[75] Derived from DoD, *Military Support to SSTR Operations JOC*, Version 2.0, December 2006, p. 60; and S/CRS, *Post-Conflict Reconstruction Essential Tasks*, April 2005, pp. IV-15 to IV-16.

[76] Derived from DoD, *Military Support to SSTR Operations JOC*, Version 2.0, December 2006, p. 60; and S/CRS, *Post-Conflict Reconstruction Essential Tasks*, April 2005, p. III-10.

[77] HQDA, FM 3-24, 2006, p. 5-15.

[78] Derived from DoD, *Military Support to SSTR Operations JOC*, Version 2.0, December 2006, p. 60.

[79] Derived from DoD, *Military Support to SSTR Operations JOC*, Version 2.0, December 2006, p. 60; and S/CRS, *Post-Conflict Reconstruction Essential Tasks*, April 2005, pp. III-8 to III-10.

[80] Derived from DoD, *Military Support to SSTR Operations JOC*, Version 2.0, December 2006, p. 60.

- **Build, restore, maintain, and operate telecommunication networks.** In the uncertain aftermath of a natural disaster, man-made disaster, or conflict, the ability to build, restore, maintain, and operate telecommunication networks.[81]

Support Economic Development

The primary goal of this MME is to promote economic development that addresses near-term problems such as large-scale unemployment and the re-establishment of economic activity in a way that lays the foundation for sustained economic growth that stimulates indigenous economic activity. A viable economy is a key component of stability and reinforces government legitimacy.[82]

- **Generate employment.** Design, fund, and implement public works initiatives to stimulate micro- and small enterprise, as well as workforce development programs that rapidly provide employment for the indigenous population.[83]
- **Develop monetary policy.** Develop mechanisms and institutions, including the ability to set and control interest rates, that allow the government to manage the economy by expanding or contracting the money supply.[84]
- **Develop and apply fiscal policy and governance.** Develop and apply sustainable, efficient, and transparent fiscal policies that can generate the resources required to sustain key public functions.[85]

---

[81] Derived from DoD, *Military Support to SSTR Operations JOC*, Version 2.0, December 2006, p. 60.

[82] Derived from the Army and DoD definitions for supporting economic development. HQDA, FM 3-25, December 2006; DoD, *Military Support to SSTR Operations JOC*, Version 2.0, December 2006, pp. 43–44.

[83] Derived from DoD, *Military Support to SSTR Operations JOC*, Version 2.0, December 2006, p. 61; the S/CRS essential tasks for this sectoral task.

[84] Derived from the Economic Growth program element 1.2, Monetary Policy, definition in *Foreign Assistance Standardized Program Structure and Definitions*, 20 October 2006.

[85] Derived from the Economic Growth program element 1.1, Fiscal Policy, definition in *Foreign Assistance Standardized Program Structure and Definitions*, 20 October 2006.

- Promote general economic policies.
- Establish, develop, regulate, and sustain a well-functioning and equitable financial sector.[86]
- Manage and control both foreign and domestic borrowing and debt.[87]
- Establish, develop, sustain, and enforce trade policies, laws, regulations, and administrative practices that support improvement in the trade environment and which facilitate international trade.[88]
- Promote a market economy. The ability to support the establishment or re-establishment of a functioning market economy.
- Promote legal and regulatory reform. The ability to support the development of a legal and regulatory framework supportive of a market economy.
- Promote agricultural development. The ability to support the establishment or re-establishment of viable agricultural sector capable of long-term growth.[89]
- Establish a social safety net. The ability to support the establishment social safety net programs.
- Build and maintain transportation infrastructure. The ability to design, execute, and sustain investment and regulatory programs that support and strengthen reliable and affordable transportation systems, including roads, airports, railways, and ports.[90]

---

[86] Derived from the Economic Growth program area 3, Financial Sector, definition in *Foreign Assistance Standardized Program Structure and Definitions*, 20 October 2006.

[87] Derived from the Economic Growth program subelement 1.2.5, Debt Management, definition in *Foreign Assistance Standardized Program Structure and Definitions*, 20 October 2006.

[88] Derived from the Economic Growth program element 2.1, Trade and Investment Enabling Environment, definition in *Foreign Assistance Standardized Program Structure and Definitions*, 20 October 2006.

[89] HQDA, *Stability Operations*, FM 3-07, October 2008, p. 3-17.

[90] Derived from the Economic Growth program element 4.3, Transport Services, definition in *Foreign Assistance Standardized Program Structure and Definitions*, 20 October 2006.

- **Develop, strengthen, and support telecommunications infrastructure through investment and regulatory reform.**[91]
- **Develop and maintain energy infrastructure.** Develop, execute, and sustain programs that increase the efficiency and reliability of energy services and which promote investment in the development, transport, processing, and utilization of indigenous energy sources and imported fuels.[92]
- **Build and maintain general infrastructure.** Develop, execute, and sustain general infrastructure programs that promote overall and municipal indigenous governance, commerce, and social well-being.[93]

Conduct Strategic Communications

The primary goal of this MME is to effectively communicate to key local and foreign audiences information regarding the stability operation in order to preserve conditions favorable to achieving the overall stability operation goals and objectives.[94]

- **Conduct public information and communication activities.** Support the objectives of the stability operation through the communication of truthful, timely, and factual unclassified information within the area of operations to foreign, domestic, and internal audiences.[95]

---

[91] Derived from the Economic Growth program element 4.2, Communications Services, definition in *Foreign Assistance Standardized Program Structure and Definitions*, 20 October 2006.

[92] Derived from the Economic Growth program element 4.1, Modern Energy Services, definition in *Foreign Assistance Standardized Program Structure and Definitions*, 20 October 2006.

[93] Derived from the S/CRS essential tasks for this sectoral task.

[94] DoD, *Military Support to SSTR Operations JOC*, Version 2.0, December 2006, pp. 61–62.

[95] Derived from the Joint Public Affairs Operations tier 1 joint capability area definition. *Joint Capability Areas Tier 1 and Supporting Tier 2 Lexicon: Post 24 August 2006 JROC*, August 2006, p. 37. An alternative definition derived from the *SSTR Operations JOC* is "the capability to conduct effective strategic communications that engage key local and foreign

Conclusion

This chapter provides an overview of the building partner capacity roles for the State Department, USAID, and DoD, and describes in greater detail the roles, missions, and capabilities for stability operations for DoD and the State Department. In particular, we have identified and briefly described the critical capabilities that are required for each major mission element. The two main areas outlined in this chapter, BPC and stability operations, indicate that there is no overarching interagency strategy for using BPC activities to build specific stability capabilities. To develop such a strategy, mechanisms for aligning Army, DoD, and national BPC for stability operations planning and resourcing will need to be constructed at different organizational levels. At the highest level, this could entail establishing a security sector assistance mechanism, jointly managed by DoD and DOS, for determining overall BPC goals and responsibilities, to include those relevant to stability operations.[96] At an intermediate level, DoD and DOS agencies in Washington and overseas could collectively determine objectives for employing and developing resources and capabilities. At a lower level, DoD and DOS security assistance planners and programmers could formulate detailed BPC for stability operations "roadmaps" for priority partner countries, perhaps within the context of Army and other DoD security cooperation and campaign planning conferences.

Although DoD must also be prepared to support civilian agencies in all six of the MME categories above, and possibly be able to conduct key tasks in all categories on its own or at least until civilian agencies are able to do so, it will probably focus its resources on the first mis-

---

audiences in order to create, strengthen, or preserve conditions favorable to the achievement of overall SSTR goals and objectives." DoD, *Military Support to SSTR Operations JOC*, Version 2.0, December 2006, p. 61.

[96] A potential model for this interagency approach is the United Kingdom's Conflict Prevention Pool, which allows the Ministry of Defence, Foreign and Commonwealth Office, and Department for International Development to collectively manage resources and funding for conflict prevention and post-conflict stabilization. See UK Department for International Development, *The Global Conflict Prevention Pool: A Joint UK Government Approach to Reducing Conflict*. As of February 2010:
http://www.dfid.gov.uk/Documents/publications/global-conflict-prevention-pool.pdf

sion element, "provide a safe and secure environment."[97] Therefore, we will use this MME as an illustrative end state in the descriptive analysis and in-depth case studies that follow in Chapters Three and Four, respectively.

---

[97] The study team chose to use the DoD term "establish and maintain a safe and secure environment" throughout this report rather than the FM 3-0 term "civil security." Both refer to the same stability mission.

CHAPTER THREE

BPC for Stability Operations Programs and Activities

Having synthesized U.S. government guidance in order to identify roles, missions, and capabilities for building partner capacity to conduct stability operations, this chapter describes the BPC for stability operations activities and programs currently being conducted by the Department of Defense, other U.S. government agencies, and major U.S. allies. This analysis will help Army leaders to better understand what BPC for stability operations programs and activities are being conducted—both within the Army and elsewhere.

This baseline analysis indicates that the U.S. Army has policy, planning, and resource management authority over only a small fraction of these activities; most are controlled and managed by other DoD components, civilian government agencies (particularly DOS), and major U.S. allies. In a capacity-building environment in which the Army is only one player among many, Army security cooperation planners and executors should attempt not only to understand what others are doing and why but also to tailor their programs and activities to ensure that they complement, or at least do not work at cross-purposes with, the efforts being pursued by other capacity-building organizations.

Although many BPC programs include different kinds of activities and events (also known as methods or "ways"), the primary activity types include

- conferences
- workshops

- information exchanges
- training
- education.

Although the lack of detailed, comprehensive, and accessible data makes even descriptive analysis difficult, it is clear that a significant number of BPC events in certain combatant commands are focused on stability operations. For example, in SOUTHCOM's case, the largest share of stability operations events is focused on building partner capacity. Stability operations activities directly executed by the United States and nonspecialized BPC activities that are useful for stability operations compose important parts of SOUTHCOM's security cooperation portfolio. Interestingly, focused BPC for stability operations programs are largely associated with capabilities different from those in "direct" stability operations events. Also, stability operations-useful events are disproportionately targeted at one country, whereas stability operations-related events are more equally distributed across multiple countries.

Most U.S. and allied efforts to build partner capacity for stability operations take place within Europe and Africa and focus on establishing a safe and secure environment through education, training, and exercises. In contrast to the United States, however, the allies favor a long-term approach to working with partners, primarily because of their cultural and colonial ties with certain countries and regions.

In an effort to provide a thorough overview of BPC for stability operations activities, this chapter first examines a few relevant U.S. Army programs before turning to select DoD and non-DoD programs. The second main section provides an overview of our analysis of the BPC for stability operations events that occurred in FYs 2005–2006 in countries within SOUTHCOM's area of responsibility, as captured by that command's Theater Security Cooperation Management Information System. This analysis provides a detailed picture of programs and activities that were conducted over a two-year period by the U.S. military in a part of the world (Latin America and the Caribbean) where capacity building and stability operations have long been important security concerns. The process outlined below of categorizing TSCMIS events related to BPC and stability operations should be helpful in establishing a baseline for future DoD security cooperation planning and programming not only in SOUTHCOM but also in other regional commands—in particular, U.S. Africa Command (AFRICOM), where U.S. security interests resemble those in SOUTHCOM. The final section of this chapter is an overview of allied stability operations programs in order to identify possible partnering opportunities and to avoid overlaps and fill existing gaps. In particular, we review programs in Australia, Canada, France, Germany, Turkey, and the United Kingdom.

U.S. Government Programs and Activities

As mentioned above, our analysis found that the Army conducts a relatively small share of all BPC for stability operations programs. In our review of 94 U.S. government programs, only 15 were Army-led programs. We cast a wide net in an effort to identify as many U.S. government programs related to BPC for stability operations as possible.[1] We discovered that many U.S. government actors are involved in BPC for stability operations around the world, including:

- Department of Defense (including the U.S. Army)
- Department of State
- U.S. Agency for International Development
- Department of Homeland Security (e.g., Coast Guard, Customs and Border Patrol)
- Department of Justice
- Department of Energy
- Department of Agriculture
- Department of Commerce
- Department of Transportation.

Although we cannot say for certain that we have identified all related BPC for stability operations programs, we have identified the

---

[1] This analysis entailed a thorough review of relevant literature, databases, and official documents, as well as focused discussions with a variety of program managers.

largest and most prominent programs conducted by the interagency actors. Examples are provided in the short sections that follow. A more complete listing can be found in Appendix B (Tables B.1–B.4).

We examined the overall objective of each *program* (also called the program "end") according to the six DoD major mission elements (as detailed in Chapter Two) and found that the majority of programs best support the "establish and maintain a safe and secure environment" MME. Table 3.1 provides a snapshot of how the programs we examined relate to each of the MMEs.

Although Table 3.1 shows that the "establish and maintain a safe and secure environment" MME is by far the most prevalent, more than half of all U.S. stability operations activities also address other MMEs, most notably "reconstruct critical infrastructure and restore essential services" and "support economic development." In particular, many USAID programs have the former as the primary MME because they address public health or development of essential infrastructure.[2]

We refer to "primary" MME focus in Table 3.1 because many of the programs and related activities and events we examined could potentially fall into more than one category. For example, a COCOM exercise on disaster response could address both the "safe and secure

Table 3.1
BPC for Stability Operations Programs Linked to MMEs

Primary Major Mission Element	Army	Other DoD	Total DoD	Other U.S. Government
Establish and maintain a safe and secure environment	6	15	21	28
Establish representative, effective governance and the rule of law	3	1	1	5
Deliver humanitarian assistance	6	2	8	3
Reconstruct critical infrastructure and restore essential services	0	0	0	11
Support economic development	0	0	0	14
Conduct strategic communications	0	0	0	0

---

[2]   Discussion with USAID, April 2007.

environment" MME but might also address the "humanitarian assistance" MME. Likewise, an Army Corps of Engineers workshop might address the MMEs of "reconstruct critical infrastructure and restore essential services" and "support economic development."

U.S. Army Programs

Of the 15 U.S. Army BPC for stability operations programs that we examined, a plurality support the "establish and maintain a safe and secure environment" MME.

Of those, one of the largest and most developed is the Army's Military Personnel Exchange Program (MPEP), which was established to foster mutual understanding between the military establishment of each participating nation by giving exchange personnel familiarity with the organization, administration, and operations of the host organization. The ranks and grades of the exchanged personnel will be equal (as much as possible), as agreed upon by the participating armies, and the types of MPEP activities range from education to joint training to human rights programs. These training and working relationships establish, on a mutually agreeable basis, an understanding and appreciation for the policies and doctrines of the respective armies. The main benefit of the exchange program is that it promotes a mutual understanding of how other militaries operate, which is invaluable when allied forces conduct operations together, especially on short notice, and are required to produce effective combat power. In addition, these exchanges reassure potential future allies and partners of America's goodwill and its strength, build professionalism and respect for rights and democracy, and open and sustain unofficial channels of communication and influence.

The U.S. Army also has a number of rule-of-law programs in place. Aside from the rule-of-law/governance programs being conducted at the brigade level throughout Iraq and Afghanistan, the following are notable Army rule-of-law initiatives:

- The West Point Center for the Rule of Law (CRL) is an academic and military center dedicated to promoting a profound respect for the rule of law during both peacetime and armed conflict.

- The Judge Advocate General's Legal Center and School (TJA-GLCS) conducts an interagency rule-of-law course for those deploying to conduct such missions.
- The Center for Law and Military Operations (CLAMO) collects lessons learned with regard to rule of law and other facets of stability operations. In addition, it publishes a Rule of Law Handbook for deploying rule-of-law practitioners.

Other Department of Defense Programs

Although the Army has the lead for many key BPC for stability operations programs, we examined a number of other relevant stability operations programs throughout DoD, most of which also support the "establish and maintain a safe and secure environment" MME.

Perhaps the most relevant DoD program supporting BPC for stability operations is the Center for Hemispheric Defense Studies (CHDS) with its education, research, outreach, and knowledge-sharing activities on defense and security issues affecting the Americas.[3] As well as fostering partnerships and advancing defense and security decisionmaking processes, CHDS offers advanced courses in stability operations to expand participants' knowledge and abilities to plan and conduct such missions. It also provides National Security Planning Workshops that allow decisionmakers in the Western Hemisphere to meet and address stability operation issues. A fairly new initiative, the Faculty Outreach Program, brings one or two faculty members to countries with active CHDS alumni associations to update alumni on the center's latest initiatives and, perhaps more importantly, to present research on major topics, including stability operations. In these ways, CHDS is a valuable tool that supports BPC for stability operations.

Other U.S. Government Programs

As Table 3.1 shows, there are also many non-DoD programs that support BPC for stability operations objectives across a greater range of primary MMEs than the DoD programs.

---

[3] CHDS's web site: http://www.ndu.edu/chds/

For example, the Export Control and Related Border Security Assistance (EXBS) program is an interagency program managed by DOS yet implemented by the departments of State, Commerce, Energy, and Homeland Security.[4] The aim of EXBS is to prevent the proliferation of weapons of mass destruction and missile delivery systems, as well as conventional weapons, by helping foreign governments establish and implement effective export control systems. The program helps countries improve their capability to prevent and interdict shipments by providing practical assistance tailored to each country's needs. For example, there are currently 20 dedicated program advisors at U.S. embassies working to coordinate and implement the EXBS program. A real advantage of this program is the broad range of countries it works with. For example, the EXBS focuses on both weapons of mass destruction "source countries" as well as states on potential smuggling routes. Similarly, EXBS works on a regional and multilateral basis in order to harmonize national export control systems with international standards and facilitate information sharing.

Another good representative example of a non-DoD program that helps build partner capacity for stability operations is the Department of Justice's International Criminal Investigative Training Assistance Program (ICITAP).[5] Although it is a Department of Justice program, DOS, USAID, DoD and the Millennium Challenge Corporation are also involved. The main goal of this program is to develop professional law enforcement institutions that protect human rights, combat corruption, and reduce the threat of transnational crime and terrorism. ICITAP achieves this goal by providing technical assistance, equipment, and training to countries worldwide in the field of forensic sciences. The program tends to design each partnership with the individual host country and to date has worked with more than 60 countries.

Appendix B includes a series of tables that provide brief synopses of each of the Army, other DoD, and other U.S. government BPC for stability operations programs that we examined in our analysis. The

---

[4] EXBS's web site: http://www.state.gov/t/isn/ecc/c27911.htm

[5] ICITAP's web site: http://www.usdoj.gov/criminal/icitap/

following section provides an overview of BPC for stability operations activities conducted by key allies around the world.

Programs and Activities in the SOUTHCOM Region

U.S. SOUTHCOM Events

In contrast to the previous qualitative analyses of U.S. government and allied activities around the world, the RAND Arroyo Center team also conducted a quantitative analysis of BPC for stability operations events in the SOUTHCOM region. Our analysis suggests ways in which event recording both inside and outside the combatant commands can be changed to improve subsequent analysis, as well as ways in which the stability operations concept itself can be expanded to more fully incorporate event data.

Using the descriptive fields provided in a SOUTHCOM TSCMIS activity report, we coded the capabilities developed in each event with respect to their applicability to future stability operations. In so doing, the stability operations concept was expanded beyond events tailored directly for stability operations to include events that could serve as stability operations building blocks. This resulted in a richer understanding of the combination of events that are being used to foster stability operations capabilities, ways in which stability operations capabilities are being constructed in individual partners, and variations in BPC for stability operations strategy throughout the region.

We obtained the data for this analysis from SOUTHCOM in April 2007. They represent most—but not all—of the DoD security cooperation events that were conducted in SOUTHCOM's area of responsibility (AOR) during FYs 2005 and 2006.

The following SOUTHCOM database variables proved relevant to our quantitative BPC for stability operations analysis:

- Event title
- Event description
- Joint Chiefs of Staff (JCS) engagement category
- Event status
- Status justification

- Country
- Total U.S. participants
- Total partner nation participants.

We used event titles and event descriptions as the major sources of BPC for stability operations information. The JCS engagement category field is a mix of security cooperation "ways"—such as combined/multinational training—and "ends," such as counternarcotics assistance. Event status describes an event's progress (planned, ongoing, completed, or cancelled), while status justification is supposed to provide the rationale for an event's cancellation. The country variable describes the location of an event, not necessarily the nationalities of event participants. Total U.S. participants and total partner nation participants contain values that range from none to several thousand. Finally, funding information was not included within the SOUTHCOM FYs 2005–2006 data file provided to us.

Coding Methodology

After initially examining and cleaning the SOUTHCOM data, we determined whether or not each event prepared the United States and partner nations for stability operations. We discovered that coding only for events that pertained directly to stability operations was too exclusive. Too many events indirectly related to stability operations were being omitted. Therefore, we sliced the stability operations category into the following pieces in order to provide a fuller, more nuanced picture of the stability operations events being planned and conducted in the SOUTHCOM AOR.

Stability operations-dedicated events satisfy one or more of the stability operations MMEs or capabilities described in Chapter Two and Appendix A.

- BPC stability operations events: partner nations obtain skills, information, etc. from the U.S. military.
- Direct stability operations events: the United States provides direct assistance to partner nations (e.g., humanitarian relief or development infrastructure).

Stability operations-useful events can be thought of as operational building blocks that may be necessary to, but not sufficient for, the development of specific stability operations capabilities.[6] This category included a wide range of events and activities, including training, exercises, simulations, and courses.

Support events generally impart skills or information deemed necessary for most kinds of military operations, not just stability operations. They do not directly address stability operations objectives or capabilities.

Not applicable to stability operations applies to SOUTHCOM events that are neither related to, nor useful for, stability operations.

The missing description field pertains to events whose titles and descriptions are missing, making it impossible to determine their objectives.

Summary of SOUTHCOM TSCMIS Analysis

Our analysis of SOUTHCOM TSCMIS data produced several broad implications. In particular, COCOM (and eventually, U.S. Army) TSCMIS datasets could be substantially improved if they were refined to include stability operations-specific coding that enables a more accurate baseline description of relevant security cooperation activities than do the current TSCMIS security cooperation categories. These more robust TSCMIS datasets could help OSD and Army decisionmakers develop a BPC strategy—both in combination with detailed activity assessments and in a systematic method for prioritizing partner nations.

However, the BPC for stability operations concept needs further refinement before such a baseline can be established. Because U.S. military doctrine historically has favored the development of general-

---

[6]   To appreciate the difference between stability operations-related and stability operations-useful events, consider the following example. A unit that is trained in light infantry tactics can be utilized "in a pinch" during a peacekeeping operation. Yet these units will lack the specific training required to respond to all peacekeeping scenarios. Therefore, we define light infantry training events as being stability operations-useful. By contrast, training events, which deal with the consequences of a man-made or natural disaster, are classified as stability operations-related because they prepare units to engage in a specific stability operations activity.

purpose over specialized forces, DoD needs to reach a consensus on a method for evaluating the significance of security cooperation activities that are somewhat useful for building stability operations capacity but not directly devoted to doing so. In addition, DoD components that manage security cooperation activities for the purpose of building partner capacity need to improve their data collection and maintenance methods. Finally, for the sake of comprehensive analysis, event descriptions should specifically include budget and manpower details as well as clearly indicate the subject matter discussed and capabilities provided.

A detailed description of our methodology and specific findings of our SOUTHCOM TSCMIS analysis can be found in a classified annex, which is available from the lead authors.

Allies' BPC for Stability Operations Activities

The following section summarizes, by country, some of the more prominent allied BPC for stability operations programs and activities. While many of these programs and activities focus on peacekeeping training, the skills developed by this type of training are similar, if not identical, to the skills developed by stability operations training. We attempt to identify possible partnering opportunities that could help the United States avoid overlaps and fill existing gaps in BPC for stability operations. We realize that political imperatives may prevent such partnerships on occasion, but at a minimum, this overview will increase the Army's and the U.S. government's visibility into relevant allied programs.[7]

This chapter examines the activities of six allies (listed in alphabetical order): Australia, Canada, France, Germany, Turkey, and the United Kingdom. These allies were chosen for three reasons. First, because they are all allies, there may be more partnering opportunities for the United States. Second, these countries execute a comparatively

---

[7] The approach to identifying allies' BPC for stability operations activities focused on a detailed review of available literature, focused discussions with U.S. Special Forces personnel who were familiar with some of the programs, and focused discussions with key allied military personnel, where possible.

large set of BPC for stability operations programs, many of which focus on preparing partner countries for United Nations (U.N.) or regional peacekeeping operations. Third, because the U.S. Army holds Army-to-Army Staff Talks with each of these countries, increased Army visibility into these activities may lead to the addition of BPC partnering issues to the Staff Talks agenda. Appendix B, Table B.5, provides a detailed list of specific allies' activities.

Australia

Australia emphasizes long-term development of armed services with the goal of building regional security and defense cooperation. Focused on the South Pacific and Southeast Asia, Australia tends to work with nations who share cultural, sociopolitical, and historical ties. Its coordinating partners often include New Zealand, Tonga, Fiji, and Papua New Guinea. In its training programs, Australia uses both conventional and unconventional training methods and includes police forces and civilian contingents of economists, development assistance specialists, and budget advisors.[8]

Australia is involved in various bilateral training exercises in the region. For example, Australia's military trains the Malaysian military, while specialized Australian counterterrorism experts work with Indonesia to coordinate security efforts. Also, Australia is especially involved in building the domestic capability of the Solomon Islands through the Regional Assistance Mission to the Solomon Islands, which includes countering ethnic conflict and land rights issues, protecting the indigenous people, and assisting the government in restoring law and order.[9]

Canada

Canada focuses on strengthening the peacekeeping capabilities of foreign armed forces with the dual goals of establishing armies that can contain and resolve internal/regional conflicts without outside assistance and also increasing the quality and quantity of troops available

---

[8] Discussions with Australian Department of Defense, April 2007.

[9] RAMSI web site: http://www.ramsi.org

for peace support missions under the aegis of an international organization and/or the United Nations.

Canada's primary training program is called the Directorate of Military Training Cooperation Programme, which involves active-duty soldiers and contractors to teach and train international forces in language, professional development, and peace support missions. Training occurs both at home—with conventional forces that conduct routine training in Canada—and abroad—with specialized forces/advisory teams being sent overseas. Canada focuses heavily on "training the trainers" by training mid- to senior-level officers, with the idea that these leaders will educate their own forces.

Although there is not an explicit regional focus, Canada continues to focus its efforts on nations in Africa, Eastern Europe, and Central and South America. This geographic consistency is primarily due to the standards countries must meet to receive training rather than a planned regional bias. As a way to encourage partner participation, Canada generously contributes to its trainees' tuition.[10]

France

France emphasizes long-term development of armed services with the goal of establishing African armies that can police, contain, and resolve state or regional conflicts effectively without outside assistance. Focused on Africa because of its previous colonial ties there, France attempts to provide its African partners with a corps of military experts capable of conceiving, preparing, and participating in peacekeeping operations under the aegis of an international organization and/or the United Nations.

France's primary training program with African nations is called RECAMP, which involves active-duty soldiers, contractors, and donors in its actions to train and organize African forces in peacekeeping methods and in equipment. France focuses heavily on training and exercises, which usually involve French trainers and African

---

[10] See National Defence and the Canadian Forces, "Directorate Military Training and Cooperation: Background." As of February 2010: http://www.forces.gc.ca/admpol/newsite/mtcpbackground-eng.html

general-purpose soldiers.[11] Each exercise consists of individual training and unit-level training, and equipping units already or soon to be engaged in peacekeeping operations.[12] Key components of each exercise include a political-military seminar (i.e., study the situation, prepare a response); a staff exercise (i.e., simulate theater and necessary military decisions to put operation into effect); and a field exercise (i.e., includes testing methods of action in real time).[13] There have been five practical training sessions to date, the first of which was held from 1996 to 1998 within the Economic Community of West African States and involved four contributor countries and four donors. It concluded with the Guidimakha exercise on the border of Senegal, Mali, and Mauritania.[14] The most recent exercise was held in 2006 in Brazzaville, Congo.[15]

Germany

Germany emphasizes long-term development of military and civilian personal for U.N. operations within the framework of international conflict prevention and crisis management. Germany works to enhance the capacity of foreign armed forces through education and training. The focus is primarily on peacekeeping skills, with the goal of preparing soldiers to participate in future international crisis deployments under the United Nations or other international organizations.

Germany and its partners (e.g., Denmark, India, Italy, Japan, Netherlands, Spain, and Switzerland) train friendly African regional

---

[11] Discussions with officials from Special Operations Command Europe, March 2007.

[12] Discussion with officials of the French Embassy to the United States, April 2007.

[13] France Diplomatie, "France in the UN System," February 21, 2006, French Foreign Ministry web site. As of December 2009:
http://www.diplomatie.gouv.fr/en/france-priorities_1/international-organizations_1100/france-in-the-un-system_3281/index.html

[14] The Guidimakha exercise occurred at the end of February 1998 and involved more than 3,500 soldiers from eight West African countries with participation of French, U.S., and British units. This exercise marked the conclusion of the first cycle of the French RECAMP practical training sessions.

[15] United Nations RECAMP Programme, "Field Peacekeeping Training," description on U.N. web site.

organizations and nations in peacekeeping and security skills.[16] Germany does this at two main training centers that are focused on the long term and are associated with the United Nations. First, the Bundeswehr U.N. Training Center is open to officers from all U.N. member countries. Working in cooperation with the Bundeswehr Center, the Center for International Peace Operations in Berlin offers a security course for civilian peace workers.[17] Second, the Kofi Annan International Peacekeeping Training Centre in Ghana focuses on the West Africa region and concentrates on enhancing the African regional organizations' capability to conduct peacekeeping training close to numerous current operations.

Turkey

Turkey educates and trains forces from around the world in both counterterrorism and peacekeeping skills. It emphasizes long-term development of armed services and fosters political and economic relations between participants. In particular, Turkey focuses on coalition training and education and provides a forum for networking opportunities.

Turkey has two main training centers, the Center for Excellence Against Terrorism (COE-DAT) and the Partnership for Peace (PfP) Training Center, both located in Ankara. At COE-DAT, contractors and active-duty troops offer specialized training in counterterrorism to military officials and civilians. Participants include nations in NATO, PfP, and Mediterranean Dialogue, among others.[18] Hosted by the Turkish Command Forces, the PfP Training Center educates foreign military personnel from PfP and NATO countries. It allows partner countries to build individual relationships with NATO, choosing their own priorities for cooperation.[19]

---

[16] Kofi Annan International Peacekeeping Training Centre web site: http://www.kaiptc.org/home

[17] Germany's Federal Ministry of Defense web site: http://www.bmvg.de/portal/a/bmvg; see "security policy."

[18] COE-DAT web site: http://www.tmmm.tsk.tr/

[19] Partnership for Peace Training Center and its relationship with NATO: http://www.nato.int/issues/pfp/index.html

United Kingdom

In its BPC for stability operations efforts, the United Kingdom emphasizes strengthening domestic order by providing protective services and building up indigenous forces with the goal of creating nations that can contain and resolve local or regional conflicts effectively. The United Kingdom focuses on providing its partners with military experts—both active-duty soldiers as well as contracting services—capable of conducting peacekeeping operations to increase the number and strength of available participants for future U.N. peace missions.[20]

The United Kingdom BPC for stability operations efforts are global and focus heavily on training and exercises, which tend to be divided into three stages: training (structured), exercise (unstructured), and operation (in effect). The United Kingdom makes a purposeful effort not to tell a country what it needs. Instead, officials ask "what do you need and how best can we deliver this training?"[21]

The United Kingdom has three major training programs:

- **British Military Advisory and Training Team** involves permanently stationed conventional forces with long-term objectives of protecting and training indigenous forces in peacekeeping methods, particularly in West Africa (especially Ghana) as well as Central and Eastern Europe.[22]
- **British Peace Support Team** is made up of specialized forces that have a short-term focus of teaching indigenous forces specific peacekeeping skills, mainly in East Africa and the Caribbean region.

---

[20] Discussion with Ministry of Defence Assistant Director in the Policy Planning directorate, March 2007.

[21] Discussions with a former British Naval Commander formerly involved with UK theater security cooperation activities, February 2007.

[22] UK Ministry of Defence Fact Sheet. As of January 2010:
http://www.mod.uk/DefenceInternet/FactSheets/OperationsFactsheets/
DefenceInAfricaBackgroundInformation.htm
    Jan Richter, Radio Praha, "British Military Mission in Vyskov Extends Focus," July 6, 2007. As of December 2009:
http://www.radio.cz/en/article/92145

- **International Military Advisory Training Team**, made up of both military and civilians, safeguards and develops the armed services as well as focuses on securing essential services, restoring vital infrastructure, and providing public order. This program is working in Sierra Leone under the auspices of Exercise Green Eagle.[23]

Summary

Overall, the majority of U.S. and allied BPC for stability operations activities take place within the U.S. European Command (EUCOM) and AFRICOM areas of responsibility. As with the United States, the allies' activities tend to focus on the goal or end state of "safe and secure environment," primarily by employing the ways of education, training, and exercises. Both the United States and its allies tend to employ contractors alongside military or civilian trainers. However, unlike the United States, the allies tend to pursue a longer-term approach for working with partners, primarily due to their historic cultural and colonial ties with select countries and regions.

Of the countries considered in this chapter, it appears that the UK and France each has train, advise, and assist (TAA) models that provide insights that could inform the U.S. approach to BPC for stability operations. All three countries view TAA and BPC as ways to favorably shape and influence the global security environment. That said, the TAA approaches differ significantly in several key areas: trainer selection, mode of deployment, training of the trainers, and career implications for the trainer. We will discuss each in turn.

Selection. The processes in the United States and the UK for selecting trainers and advisors from the conventional forces do not appear to be particularly rigorous. Nor are these assignments generally sought out by officers in these two countries. The French model ties the selection process to career progression. Advisory duty in the

---

[23] British Ministry of Defence, Defence News, "Sierra Leone Deployment Provides Valuable Lessons for Royal Navy Amphibious Task Group." As of December 2009:
http://www.mod.uk/defenceinternet/defencenews/trainingandadventure/sierraleone
deploymentprovidesvaluablelessonsforroyalnavyamphibioustaskgroupvideoaudiopart1.htm

French army is an expectation for those officers who are competitive for advancement.

Deployments. France and the UK have similar TAA models—advisors are embedded with the partner and often wear the host nation uniform. The United States does not typically embed its advisors, although this has been the practice in Afghanistan and Iraq. Moreover, France has more of a regional approach to its TAA deployments, with Africa being the focus. The UK has a global approach, similar to the United States, but with many fewer deployments.

Training. The U.S. system for preparing trainers and advisors emphasizes operational and tactical training over "cultural" training, and what cultural training there is does not address key points, e.g., such as "empathy with the advised," as do the French and British models. Although the French and UK predeployment training for advisors lasts only about two weeks, the selection process appears to do a good job of ensuring that the "right" people are being trained. In the U.S. system, training for TAA varies from two to six months, with the selection process not as rigorous, as compared to the UK and France.

Career Implications. There are no foreign area officer programs in France and the UK; most of those deployed on TAA missions are from the general purpose forces and generalists. In the French system, the TAA mission is part of a deployed battalion's normal mission. Furthermore, advisory duty is part of the normal career path, and success on TAA missions is seen as a prerequisite for advancement. This is not the case in the UK or the United States. In the UK, TAA missions are encouraged but not necessarily career enhancing. In the United States, TAA missions have traditionally not been part of mainstream career paths. Indeed, advisory duty has generally been viewed as detrimental to advancement—it was what happened to an officer who was not competitive for more important, career-enhancing assignments. Clearly, the importance of training the military forces of Iraq and Afghanistan as components of a successful strategy is understood.

Conclusion

In this chapter the study team argues that, in order to construct a BPC for stability operations strategy, the U.S. Army must first develop a comprehensive understanding of the programs and activities designed to build the capacity of priority U.S. partners to conduct stability operations. The Army is only one relatively small player in this joint, interagency, and multinational capacity-building enterprise. Thus the Army must know what other services, U.S. civilian agencies (such as USAID and the State Department), and major allies (such as France and the United Kingdom) are doing to build stability operations capacity. With such visibility, the Army can make efforts to coordinate with these other organizations or at least not detract from their activities.

In addition, the Army needs to be thoughtful in how it accounts for security cooperation activities that relate to stability operations. The U.S. military's largest repositories of security cooperation events, the Theater Security Cooperation Management Information Systems, which are managed by each of the regional COCOMs, were not designed to provide aggregate data on particular operational activities. Therefore, Army analysts who want to establish a baseline understanding of U.S. military resources devoted to BPC for stability operations should carefully parse the TSCMIS data. They must ensure that they distinguish between events that are examples of U.S. stability operations activities and those that are truly related to building partner capacity. In addition, they must identify the events that focus on stability operations per se and those that focus on providing training and equipment that are useful for, but not essential to, stability operations.

CHAPTER FOUR

Assessing BPC for Stability Operations Programs and Activities

Building on the role, missions, and capabilities synthesis in Chapter Two and the baseline programmatic analysis in Chapter Three, this chapter provides a preliminary assessment of a range of BPC for stability operations programs. At the heart of this analysis is a six-step assessment approach designed to enable the Army and other DoD agencies to make more informed decisions about BPC for stability operations planning, programming, and budgeting. This approach provides a systematic method to evaluate existing security cooperation program and activity performance and effectiveness with respect to stability-related objectives and end states in particular countries. This approach is described in detail in the first section of this chapter.

To test this assessment approach, we analyzed six BPC for stability operations cases. Three of these are examined in this chapter, with a detailed description of the analyses and findings provided in Appendix D. Recognizing the limitations of a case study methodology that relies on a small sample set, we focused on prominent BPC for stability operations programs representing five different methods or "ways" of using security cooperation to contribute to the stability operations goal most relevant to DoD, specifically, "establishing a safe and secure environment." These methods included: training, exercises, education, defense and military contacts, and conferences/workshops.

Illustrating the Assessment Approach

The six-step approach to assess the effectiveness of a BPC for stability operations program will help the Army better understand how individual programs and specific activities contribute to the achievement of BPC for stability operations end states. It is depicted in Figure 4.1 and described in the sections that follow.

Step 1: Select Desired End State and Specific Goals

In the first step in the assessment, the assessor selects a stability operations end state and disaggregates it into its subordinate goals. We first considered DoD's list of six BPC for stability operations major mission elements (as discussed in Chapter Two and Appendix A) and selected "establish and maintain a safe and secure environment" as our illustrative end state because it most closely aligns with DoD's BPC for stability

Figure 4.1
Six-Step Approach to Assess the Effectiveness of BPC for Stability Operations

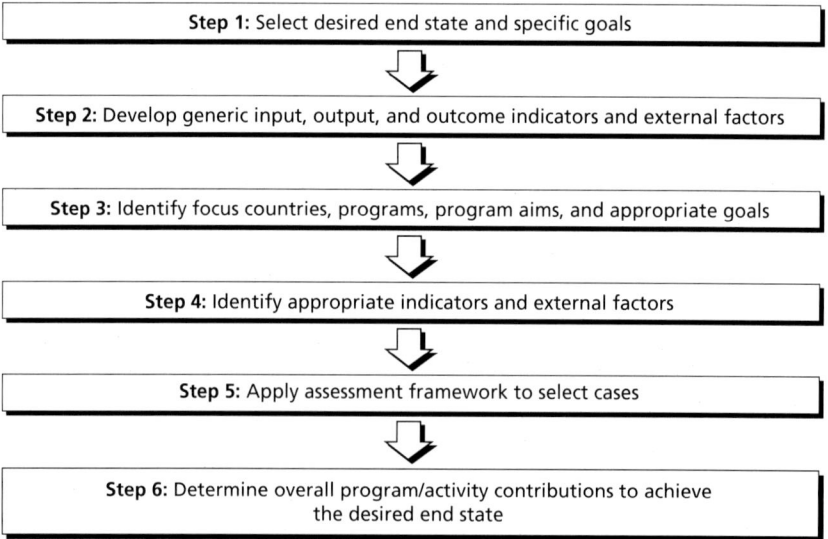

Step 1: Select desired end state and specific goals

Step 2: Develop generic input, output, and outcome indicators and external factors

Step 3: Identify focus countries, programs, program aims, and appropriate goals

Step 4: Identify appropriate indicators and external factors

Step 5: Apply assessment framework to select cases

Step 6: Determine overall program/activity contributions to achieve the desired end state

RAND *MG942-4.1*

operations role. Because the end state is so broad and, as such, not easily measurable, we disaggregated the end state into specific goals that we determined to be the most appropriate for the Army's BPC for stability operations role.[1] The end state of "safe and secure environment" comprises 12 goals, of which nine (below, in bold) correspond most closely to Army missions and are therefore potentially relevant to the Army's ability to build partner capacity for stability operations.[2]

- **Develop and enhance capability to conduct peace operations.**
- **Conduct disarmament, demobilization, and reintegration operations.**
- **Develop and sustain armed services and intelligence forces.**
- **Establish and maintain border and boundary control.**
- **Establish identification regime.**
- Provide interim public order.
- Conduct civilian police operations.
- Provide protective services.
- **Protect critical installations and facilities.**
- **Protect reconstruction and stabilization personnel.**
- **Coordinate indigenous and international security forces and intelligence support.**
- **Enable participation in stability operations-related regional security arrangements.**

We considered Army Title 10 responsibilities, Army functions, and State Department definitions to make this determination.

---

[1] The Arroyo study team determined that these goals or capabilities were easier to measure and relatively easy to link to end states such as "internal security" or "freedom from external threats." In the absence of U.S. agreed-upon standard measures for internal security or stability, the Arroyo team chose to rely on the goals laid out by DOS, which have been largely adopted in the stability operations community.

[2] The goal "provide interim public order" could, in some circumstances, be the responsibility of the Army. But we believe this goal to be more closely associated with a civilian police function.

Step 2: Develop Generic Input, Output, and Outcome Indicators and External Factors

Second, we developed generic input, output, and outcome indicators that aligned BPC ways (e.g., education, training, exercises, workshops) with stability operations goals (e.g., "safe and secure environment"). *Inputs* measure the resources, such as manpower and money, which are applied to a particular program or activity. *Outputs* measure the direct results of activities. At the most basic level, outputs help to create a baseline describing the level and type of engagement with a partner country. An example of an output indicator for training would be the quantity of forces trained for deployment.

Over time, the outputs produce *outcomes*, which measure the longer-term results of activities. An example of an outcome indicator for training would be the number of forces deployed to a specific operation. The indicators used in the regional/coalition and indigenous case study analyses below are tailored to the specific cases, and Appendix C includes a complete list of generic input, output, and outcome indicators.

We also identified external factors that impeded or facilitated the success of the program/activity in the achievement of the desired end state. These include process factors, other security ways that contribute to the end state, and specific country factors.

We selected the output and outcome indicators that best applied to each of the respective case studies. For example, if a case focuses on training, and specifically on building indigenous capacity, then appropriate output indicators include skills acquired, level of interoperability (common standards), and number of soldiers or units trained. An appropriate outcome indicator would depend on whether or not training has been institutionalized.

Step 3: Identify Focus Countries, Programs, Program Aims, and Appropriate Goals

Third, the study team identified focus countries, programs, program aims, and appropriate stability operations goals for the analysis. To do this, the team selected six BPC for stability operations case studies: three regional/coalition cases and three indigenous cases. We labeled a case study regional/coalition if it primarily focuses on multilateral capacity-building efforts, or if the goal of the program is to deploy to regional or coalition operations. Likewise, we placed a case study in the indigenous category if it primarily focuses on domestic capacity-building efforts through bilateral security cooperation ways. We selected the following countries to serve as the context for our case studies because of the relative abundance of pertinent data on them: Botswana, Cameroon, Ecuador, El Salvador, India, Jordan, Kenya, Nigeria, Romania, Rwanda, and Senegal. The programs we chose for our three regional/coalition cases and three indigenous cases are:

- **Regional/coalition cases**
  - Civil Military Emergency Preparedness (CMEP) program (examined in this chapter).
  - DoD regional center stability operations courses and conferences (examined in this chapter).
  - Peacekeeping operations (PKO) exercises in Latin America.

- **Indigenous cases**
  - African Contingency Operations Training and Assistance (ACOTA) program (examined in this chapter).
  - The Italian Center of Excellence for Stability Police Units (CoESPU).
  - State Partnership Program (SPP) in Latin America.

The six case studies represent an illustrative cross-section of BPC for stability operations activities. The study team selected case studies based on the following criteria, not in priority order:

- Security cooperation methods or ways, such as training, exercises, education, etc.
- Programs the Army controls (i.e., has policy and/or resource oversight) and those the Army does not control.
- Training method.
- COCOM area of responsibility.

Analyzing different security cooperation ways allows for consideration of a broader set of activities than the Army uses in BPC for stability operations. Consideration of programs the Army controls as well as the programs it does not control allows for greater visibility into and assessment of ongoing BPC for stability operations programs. Our analysis of training in both bilateral and multilateral cases and an illustrative case study from at least two COCOMs provides some geopolitical diversity to our assessment.[3]

Next, we identified the primary aims of each of the case studies/ programs and then determined which goals from the "safe and secure" end state were relevant to our case studies. Of the nine goals deemed appropriate for the Army (as identified above in Step 1), we concluded that the aims of our six case studies collectively support the four listed below:

- Develop/enhance capability to conduct peace operations.
- Develop and sustain armed services and intelligence forces.
- Establish and maintain border and boundary control.
- Enable participation in stability operations-related regional security arrangements.

This matrix of case study program aims and select "safe and secure" goals is shown in Table 4.1.

It is quite possible that other programs not reviewed by the study team address the other objectives. Even if this is not the case, it is certainly possible that they could be addressed through future Army BPC for stability operations activities.

Step 4: Identify Appropriate Indicators and External Factors

Fourth, since not all of our generic indicators and external factors were appropriate for each case study, we selected those that were applicable within the context of particular focus countries and capacity-building programs. For example, in some cases, foreign training programs have

---

[3]   While not an exhaustive representation of all possible cases, this sample set reflects the study team's best effort to identify a reasonable cross-section of typical BPC for stability operations cases around the world.

Table 4.1
Stability Operations Goals in Relation to Case Study Aims

Case Study Aims	Select Goals for the "Safe and Secure" End State			
	Conduct Peace Operations	Develop and Sustain Armed Services	Improve Border Control	Participate in Regional Stability Operations Arrangements
ACOTA: increase partner capacity to conduct peace support operations	X	X		X
CoESPU: train stability police unit instructors	X			X
SPP: promote regional stability and civil-military relationships		X	X	X
CMEP: improve trans-boundary cooperation on emergency preparedness				X
Regional center: provide education opportunities to BPC for stability operations				X
PKO: generate additional peacekeeping units for U.N. operations	X	X		X

involved partner units that were subsequently employed or deployed in support of stability operations. But this has not always occurred, and even when it has, the U.S. government often has not had the capability to evaluate the operational results of these units' employment/deployment. Thus it is not practical in every case to assess training outcomes in terms of the effectiveness of unit capabilities employed or deployed for stability operations.

Even more than indicators, external factors can be quite case-specific. Each partner country is likely to differ somewhat from others in terms of the security cooperation "package" that it has received from the United States and other "security exporters." Furthermore, the social, political, and economic environment in which U.S. capacity-building programs operate is different in every partner country. Consequently, the specific external factors that should be considered when evaluating the conduct of BPC for stability operations programs can only be determined within the context of the individual case study.

Step 5: Apply Assessment Framework to Select Cases

Fifth, the study team conducted assessments of each case study to determine whether the program is producing the desired outputs and outcomes in the selected countries. We obtained evidence from a variety of security cooperation sources: unclassified reports, after-action reviews, program and activity assessments, and programs of instruction produced by the U.S. Army and other DoD organizations, as well as focused discussions with key U.S. policy planners and program managers, program and activity executors, and partner country officials.[4]

For each case study, we first determined the inputs for the specific activity—specifically the funding and manpower necessary to execute the activity. Next, we identified the appropriate output and outcome indicators, and applied the available data about the select activity against these indicators. Because the end state (in this case, "safe and secure environment") is rather broad, it is necessary to focus the analysis on one or more specific goals. Sometimes external factors outside of U.S. control have to be taken into account in the overall assessment of how well an activity is meeting the end state. This analysis will be illustrated in greater detail in the three case studies later in this chapter.

Step 6: Determine Overall Program/Activity Contributions to Achieve the Desired End State

Sixth, the study team assessed how successful each program/activity was at achieving the desired end state—a safe and secure environment—as well as each of the four specific objectives identified in Step 1. This was accomplished by first summarizing the data we collected on program inputs, outputs, and outcomes and then accounting for factors that may have impinged on program results but were outside the U.S. military's control. We then subjectively balanced these internal and external fac-

---

[4] In three of the cases, we spoke with partners. These include CMEP, ACOTA, and the regional centers.

tors and arrived at a tentative conclusion regarding overall program success in our focus countries at the time that our research was conducted. These conclusions cannot be more definitive because of the limited data available to us and the prolonged gestation period for some of these programs. In addition, the ultimate success or failure of these programs will likely result from the interaction of a number of factors, only some of which can be predicted or affected by U.S. officials.

The next section of this chapter focuses on the application of the assessment framework, or Step 5 of the above approach.

Applying the Assessment Framework: BPC for Stability Operations Case Studies

This section describes three of the six BPC for stability operations case studies that we conducted. First, we will examine the Civil-Military Emergency Preparedness Program and the regional centers (specifically the George C. Marshall Center), both examples of regional/coalition capacity cases, and then we will detail the Africa Contingency Operations Training Assistance Program, which is an example of an indigenous capacity case. Within each case study, a general description of each case is presented, followed by a discussion of each of the four components that are integral to the assessment that is conducted in Step 5 of the six-step approach: inputs, outputs, outcomes, and external factors. A detailed description of the analyses and findings for each of these three cases is provided in Appendix D.

Civil-Military Emergency Preparedness Program

Case Study Overview. The main aim of the U.S. Army's International CMEP is to encourage transboundary cooperation on emergency preparedness among countries that participate in NATO's Partnership for Peace program through joint disaster preparedness exercises. CMEP is overseen by the OSD/Partnership Strategy office and the HQDA G-35 office, and implemented by the U.S. Army Corps of Engineers. CMEP contributes to the end state of "safe and secure environment" by conducting exercises. The study team conducted research on CMEP table top exercises (TTXs) through a review of relevant strategy documents, briefings, and focused discussions with U.S. and partner officials. Additionally, the study team participated in a week-long CMEP TTX, ALBATROSS 2007, in Batumi, Georgia in February 2007. The team focused its assessment on Romania, a long-standing CMEP member.[5]

Inputs. Typical input indicators used for the regional/coalition cases include money and manpower. In particular, CMEP is funded by the Warsaw Initiative Fund, overseen by OSD/Partnership Strategy. The CMEP budget has steadily decreased, from $3.1 million in 2005, to $2.3 million in 2006, to $1.2 million in 2007. The average cost of a CMEP TTX is $350,000–$400,000.

Manpower requirements for regional/coalition case studies tend to vary greatly, with CMEP events requiring only about ten U.S. officials, on average, while a typical PKO exercise may include about 200–500 U.S. and partner participants. In contrast, DoD regional center stability operations courses tend to include one to two main instructors and a small number of supporting adjunct staff.

Outputs. As with the indigenous cases, regional/coalition output indicators focus on frequency- and quantity-related issues. For some cases, this means quantity of exercises or events held each year, or whether the partner sent the appropriate number of individuals at the right level of experience and sufficient quantity to take part in the event. An additional output indicator applied to CMEP exercises considers whether operational or technical problems were identified.

Regarding quantity of exercises held, CMEP typically facilitates one major multinational TTX per year in a partner nation, on a rotating basis. With CMEP, the capabilities exercised are dependent upon the interests of the host nation. For example, Romania chose to test its new civil defense structure and exercised existing geographic information system capabilities at the TOMIS TTX in 2005.

---

[5] For our analysis, the team spoke with representatives from the Romanian Ministry of General Inspectorate for Emergency Situations in the Ministry of Interior who have long participated in CMEP events.

The regional/coalition cases generally attract the appropriate number and type of representatives from the partner countries. In fact, some of the same officers and officials tend to return year after year to CMEP exercises.[6]

Outcomes. Outcome indicators for the regional/coalition cases focused more on follow-on types of occurrences, such as whether common operational and technical problems were resolved, whether common standards were adopted, if capabilities had been deployed, and if relations built in the earlier phases of a BPC effort were being maintained through some other venue.

Especially regarding exercises, it is important to note if the operational and technical problems identified during a specific exercise event were resolved. For example, following CMEP exercises, Romania developed a U.S.-influenced perspective of civil-military and transboundary cooperation for disaster preparedness and response. The Romanian Ministry of General Inspectorate for Emergency Situations then sought and obtained a loan from the United Nations to enhance its emergency management capabilities (i.e., training its forces) to improve earthquake response capabilities. During the 2005 TOMIS exercise, Romania tested the response capabilities of its new response plan, and this system was apparently used in response to an outbreak of bird flu the following year. During the 2007 Batumi TTX, Romania proposed that an independent source be invited to assess the civil emergency preparedness systems of all CMEP countries.[7]

External Factors. It is worth noting two points about CMEP external factors. First, additional security cooperation ways have helped CMEP to succeed in Southeast Europe, in particular. Mutually supporting regional initiatives such as Stability Pact, NATO's Euro-Atlantic Disaster Response Coordination Centre and Civil Emergency Planning Directorate, and the Black Sea Initiative all enable CMEP success.

---

[6] CMEP data based on discussions with CMEP program managers and discussions with partner officials during a CMEP TTX.

[7] This assessment has since occurred, albeit with fewer countries than anticipated. Phase I is a vulnerability and needs assessment; Phase 2 is an implementation plan for upgrading systems.

Second, in terms of specific country factors, Romania is suffering from poor economic conditions including low gross domestic product and high national debt. Although Romania remains an enthusiastic CMEP partner, economic conditions are an inhibitor to implementing its robust reform agenda.

CMEP and Romania. In this scenario, we considered how well CMEP, as conducted in Romania, contributes to the objective of "participate in regional stability operations arrangements" through education and exercises.

As shown in Table 4.2, although CMEP struggles with limited inputs, program manager collaboration with other regional initiatives has partially offset this limitation. The result is a greater ability to achieve its aims, as demonstrated by Romania's positive output and outcome results.

DoD Regional Centers Stability Operations Courses and Conferences

Case Study Overview. The Department of Defense operates five regional centers that aim to provide educational courses designed to build partner countries' capacity to contribute to stability operations. We considered how each approaches stability operations within its courses, conferences, and other outreach activities. The centers contribute to

Table 4.2
Summary of CMEP in Romania Relative to Indicators and External Factors

Inputs	• Steady budget has been cut • Limited manpower available
Outputs	• Exercised new civil defense structure and geographic information system capabilities • Identified operational and technical problems
Outcomes	• Obtained a U.N. loan to enhance transboundary emergency preparedness capabilities • Tested capabilities during CMEP TTXs
External factors	• Cooperation with other regional initiatives enhanced CMEP's ability to achieve its aims

the end state of "safe and secure environment" through the methods (ways) of education and conferences in the respective regions.

Four of the centers have stability-related courses: the George C. Marshall Center in Garmisch, Germany; the Asia-Pacific Center for Strategic Studies in Honolulu, Hawaii; the Near East South Asia Center for Strategic Studies, which is part of the National Defense University (NDU) at Fort McNair in Washington, D.C.; and the Center for Hemispheric Defense Studies, also part of NDU at Fort McNair. The fifth school, the Africa Center for Strategic Studies, in NDU at Fort McNair, does not offer stability operations courses, but focuses instead on general leadership programs.

For our analysis, the study team focused on the George C. Marshall Center and Romania. The Marshall Center's mission is to create a more stable security environment by

- Advancing democratic defense institutions and relationships.
- Promoting active, peaceful engagement.
- Enhancing enduring partnerships among the nations of America, Europe, and Eurasia.[8]

We conducted our analysis through a detailed review of the programs of instruction as well as focused discussions with course leaders and other academic and administrative staff at the Marshall Center. We focused our analysis on Romania in terms of its participation in stability operations-related courses and conferences offered by the Marshall Center.[9]

Inputs. The DoD regional centers receive their resources from the OSD Warsaw Initiative Fund (Marshall Center only), the Counterterrorism Fellowship Program, and operations and maintenance resources. Some centers have recently received increases in funding (e.g., Near East South Asia Center, Africa Center, and the Center for

---

[8]   From the Marshall Center's Mission Statement, available on its web site: http://www.marshallcenter.org/mcpublicweb/en/nav-mc-about-mission.html

[9]   We spoke with Romanian officials who have participated in Marshall Center conferences (vice courses).

Hemispheric Defense Studies), while the larger centers (the Marshall Center and the Asia-Pacific Center) have seen their budgets slightly cut. It is worth noting that none of the centers have received additional resources for stability operations-related courses. As with CMEP, the regional center resources are monitored by OSD.

For manpower, the regional center stability operations courses tend to include one or two main instructors and a small number of supporting adjunct staff.

Outputs. The Marshall Center offers students a host of different elective courses, of which students can select three subjects to study during their time in the program. One of those courses is entitled "Program for Peace Support and Stability Operations." This elective three-week course commenced in 2004 and is taught three times per year. The course consists of four thematic modules featuring presentations by expert U.S. and international civilian, military, and government practitioners. Each module is followed by small group seminars, in which participants debate and exchange ideas on the issues presented.[10] In addition, several case studies, exercises, and an extended field trip serve to reinforce the topics discussed.[11]

In our examination of Romanian attendance at the Marshall Center, we learned that representatives from the ministries of Defense, Interior, and Foreign Affairs regularly attend Marshall Center courses every year.

Based on our observations during six separate Marshall Center conferences held in Croatia, Germany, Macedonia, Bulgaria, Lithuania, and Montenegro in 2005–2007, as well as discussions with Marshall Center and partner country representatives, Romania sends appropriate representatives to events in relatively large numbers, and occasionally funds the travel of its participants. Attendance varies according to topic, but in general, government officials and NGOs from a variety of sectors regularly attend these conferences. Romania

---

[10] Module I: General Peacekeeping; Module II: Security and Stability; Module III: Transition and Reconstruction; Module IV: Capacity Building.

[11] Discussions with course leaders, February and May 2007, Garmisch, Germany.

also provided five key speakers to Marshall Center conferences from 2005 to 2007.

Outcomes. In terms of maintaining relations and creating opportunities for follow-on arrangements, the activities of regional centers are worth highlighting. The Marshall Center maintains a robust alumni network, and Romania has one of the most dynamic alumni programs among all partners. This is evidenced by Romania's ability to both organize and fund conferences on key strategic issues, as well as to stay firmly connected to the Marshall Center. For example, the Romanian alumni program organized a high-level conference in October 2006 on energy security, which the president of Romania and other high-level officials attended.

External Factors. Of the three types of external factors (i.e., process factors, other security cooperation ways, and country factors) that could either hinder or facilitate the success of BPC efforts, a few points are worth noting. First, on process factors, the regional centers could greatly benefit from increased coordination on stability operations-related courses, conferences, and outreach events. Currently, each center runs its own independent courses with little interaction among the various course leaders on the subject. It is the study team's assessment that they may be missing opportunities to collaborate and share best practices among the centers.

The Marshall Center and Romania. As with the CMEP scenario, we considered how well the Marshall Center contributes to the objective of "participate in regional stability operations arrangements" for Romania.

As shown in Table 4.3, the Marshall Center seems to be achieving its aim in Romania, although the results are not very tangible. Its effectiveness in BPC for stability operations could potentially be improved by greater collaboration with other schools and centers in the region.

Because both CMEP and the Marshall Center appear to be achieving their aims in Romania, the overall effect seems to be a fairly significant contribution to the overall achievement of the objective: "participating in regional stability operations arrangements."

Table 4.3
Summary of the Marshall Center and Romania Relative to Indicators and External Factors

Inputs	• Budget cut, no additional funding for stability operations courses • Adequate personnel for stability operations-related events
Outputs	• Regularly attends courses, conferences • Provides key speakers • Often funds own way
Outcomes	• Maintains a robust alumni network that organizes and funds conferences on key strategic issues
External factors	• Lack of collaboration on stability operations with other regional centers may mean missed opportunities

Africa Contingency Operations Training Assistance Program

Case Study Overview. The ACOTA program provides training to African militaries in order to increase partner countries' capacity to conduct peace support operations. Key aims of ACOTA include imparting peace support operations skills for troops and battalion-level command staff, and building and sustaining partner countries' capacity to train their own peace support forces. ACOTA is executed under the auspices of the Global Peace Operations Initiative (GPOI), which was established after the 2004 G-8 summit meeting to address growing gaps in international peace operations,[12] and is the successor program to African Contingency Response Initiative (ACRI). While ACRI emphasized training in nonlethal peacekeeping skills, ACOTA emphasizes capability sustainment through training the trainers. The State Department's Bureau of Political-Military Affairs oversees GPOI and manages ACOTA funding, and the DOS Bureau of African Affairs executes the ACOTA program in collaboration with OSD Afri-

---

[12] For more information on GPOI, see Nina Serafino, "The Global Peace Operations Initiative: Background and Issues for Congress," CRS Report RL32772, updated March 19, 2009.

can Affairs. An Interagency Policy Development Oversight Committee provides high-level direction for ACOTA.[13]

ACOTA uses training to build indigenous partner capacity, which helps meet the desired end state of a safe and secure environment in Africa. To conduct this assessment, the Arroyo study team reviewed U.S. government documents, U.N. documents, and scholarly articles, and spoke extensively with U.S. and foreign officials closely involved with ACOTA, both on the planning and execution sides.

Inputs. Typical input indicators we used for the indigenous cases include money and manpower. For the ACOTA case we also considered a third input indicator, capabilities trained.

As mentioned above, ACOTA is funded through the GPOI program, which is overseen by DOS. ACOTA spends about $1 million on equipment for each battalion trained, depending on an assessment of need. The average ACOTA investment per soldier trained and equipped entirely by the United States is roughly $3,700.[14] The State Department provides funding for lethal equipment on a case-by-case basis to African countries prior to deployment as part of GPOI funding that is separate from ACOTA.[15] However, U.S. and African officials noted that deployed African battalions are often ill-equipped and could benefit from additional training using the lethal equipment that they use on their missions.

In terms of manpower for ACOTA, the United States sends a training team of roughly 15 to 20 contractor trainers to each country until a partner has attained the skills to independently train its military for peace support operations. The contractors typically are retired U.S.

---

[13] For example, the United States suspended ACOTA training to Uganda following Uganda's invasion of the Democratic Republic of the Congo in 1998. The United States resumed the partnership in 2007.

[14] This figure is a study team calculation based on State Department estimates: DOS officials said that $1.2 million is a rough estimate of the cost to train one partner country battalion (in cases in which no partner trainers are contributing), and $1 million to equip one battalion. We assume a 600-person battalion.

[15] According to a senior OSD official, DoD does not provide equipment because it cannot do so unless the recipient country has signed an Article 98 agreement, which is a bilateral nonsurrender agreement of U.S. citizens to the International Criminal Court.

military enlisted and officers, along with three to six uniformed officer mentors.[16]

Outputs. Output indicators for the indigenous cases focus on frequency- and quantity-related issues, such as the number of soldiers or units trained. For others this means number of events or percent of stability operations-related events per year. The other common output indicator is whether the partner sent the appropriate number of individuals at the right level of experience and in sufficient quantity to take part in the event.

The number of host country trainers trained by ACOTA increased from 275 in FY 2005 to 909 in FY 2006. ACOTA program officials report that roughly 9,000 troops were trained in FY 2004, whereas roughly 15,000 troops were trained in FY 2006.

The second output indicator focuses on whether the partner sent the appropriate representation to BPC events. In the case of ACOTA, one area of concern is that partners sometimes send composite battalions to ACOTA training, that is, small pieces of disparate battalions who come together as a new unit for the first time during the ACOTA training. Officials cited concern that training such composite units limits the unit cohesiveness of the battalions when they eventually deploy.

Outcomes. Outcome indicators for the indigenous cases focused on more qualitative issues, including quality of soldiers/trainers/units trained; whether deployable/employable capabilities resulted from the assistance; and the types of follow-on events that resulted. The ACOTA program supported two of these goals.

In our discussions about ACOTA, U.S. and African officials as well as nongovernmental experts all cited the usefulness of the ACOTA program in improving partner capability for peacekeeping operations. Most indicated that they perceived the program to improve significantly the professionalism and skill level of the units that received training. In particular, ACOTA-trained troops also were more likely to comply with rules of engagement when confronted with hostile forces.

---

[16] This responsibility will soon be transferred to AFRICOM.

ACOTA also has had success regarding the deployment of trained units. In FY 2006, 79 percent of all African battalions deployed on peace support operations globally have received ACOTA training.[17] All ACOTA countries, with the exception of Botswana, have deployed to peacekeeping operations. All 12 of the Senegalese battalions trained by ACOTA since FY 2005 have deployed in peacekeeping missions in Côte d'Ivoire, the Democratic Republic of the Congo, Liberia, and Sudan.

Perhaps not surprisingly, there are variations in the quality of ACOTA-trained units deployed in the field. Nigerian troops tend to fall far short of expectations for effective peacekeeping, and are not as skilled as Senegalese or Rwandan troops. However, according to U.S. military observers, ACOTA-trained units in the field behave more professionally than non-ACOTA trained units.

External Factors. With respect to ACOTA, political and economic conditions have in some ways hindered the success of the program in many countries. For example, Botswana is hesitant to deploy troops on peacekeeping missions, even though it is wealthy compared to other African nations. For Botswana, then, the economic motive that creates incentives for many African countries to participate in peacekeeping operations is absent. With respect to Nigeria, U.S. observers to the African Union mission in Darfur cited a generally low level of military capacity compared to Rwandan and Senegalese troops serving there, and largely attributed this difference to historical factors that have eroded Nigerian military institutions. Rwanda's high military capability, on the other hand, may be due in part to Rwanda's tragic recent history of genocide. Senegal's relatively stable and democratic history has made it a recipient of extensive aid from donor organizations. DoD officials speculated that this favored aid recipient status may have led to a diminished resolve within the Senegalese government to become self-sufficient.

---

[17] State Department, *FY 2006 Performance and Accountability Report, Performance Section*, p. 71. As of December 2009:
http://www.usaid.gov/policy/par06/USAID_PAR06_Performance.pdf

ACOTA and Senegal. The Arroyo team conducted a number of country-specific scenarios to illustrate how assessing the achievement of a program's aims in a particular country can be tied to the achievement of a stability operations end state. In this scenario we considered how well the ACOTA program, as conducted in Senegal, supports aims compatible with our stability operations "safe and secure environment" end state. As detailed earlier in this chapter in Table 4.1, ACOTA contributes to the goals of "conduct peace operations," "participate in regional stability operations arrangements through the way of training," and "develop/sustain armed forces."

As with the other case studies, we can at most draw preliminary conclusions regarding how well the programs are achieving their aims—much less how they are contributing to the achievement of the applicable goals. As shown in Table 4.4, ACOTA seems to have a fairly substantial impact in Senegal as far as achieving its aim of increasing partner capacity to conduct peace support operations.

Although ACOTA seems to be meeting its aim, additional training with cohesive, rather than composite, units could increase Senegal's ability to participate in regional stability operations arrangements.

Table 4.4
Summary of ACOTA in Senegal Relative to Indicators and External Factors

Inputs	• Could benefit from additional training
	• Resources are used for both training and equipment
	• Adequate manpower
Outputs	• Number of troops trained increased substantially during FYs 2005–2006
	• The training of composite battalions limits unit cohesion during deployments
Outcomes	• All 12 of the Senegalese battalions trained by ACOTA since FY 2005 have deployed in peacekeeping missions in Côte d'Ivoire, the Democratic Republic of the Congo, Liberia, and Sudan
External factors	• Extensive aid from donor organizations may have led to diminished resolve to become self-sufficient

Conclusion

The six-step analytical approach outlined in this chapter can help the Army—and other U.S. government agencies—assess the performance of programs and activities in achieving BPC for stability operations goals in particular countries. However, this approach is best undertaken systematically, not in an ad hoc fashion through selected interviews and available documentary materials. Ideally, information on inputs, outputs, outcomes, and external factors should be collected regularly, widely, and over a sustained period. Otherwise, assessment results will continue to be impressionistic and the linkages among factors will be difficult to determine.

To effectively implement an assessment framework such as the one described above will require extensive coordination among various DoD agencies with respect to data collection, aggregation, integration, and analysis. Such coordination could be facilitated by existing Army and other DoD security cooperation forums as well as by COCOM and Army-managed security cooperation information systems. But these gatherings and data management tools will have to be reconfigured to support systematic assessment. Moreover, if programs and activities are ever to be assessed with respect to their relative contributions to achieving BPC end states, performance measures will have to be selected in a way that balances specificity and comparability. Also, assessors must keep in mind that DoD security cooperation programs are only one (usually small) element in the overall BPC equation. Thus knowing with certainty that this or that program has achieved a long-term impact in a specific country will generally not be possible. And even if it were possible, external factors beyond the control of security cooperation officials may obviate the achievements of even well-conducted programs.

Finally, performance assessment is designed to provide program managers and policymakers with an indication of how well things are going in a partner country. It does not directly address whether the U.S. government should be expending resources trying to build partner capacity in country X versus country Y. A different kind of analytical process is needed to select appropriate capacity-building partners for different types of stability operations. The next chapter describes such a process.

CHAPTER FIVE

Analyzing Potential Partners

In an effort to provide some analytical rigor and standardization to the partner-selection approach, this chapter proposes a relatively simple spreadsheet method for determining potential partners, pros and cons of each partner, and ways to weight and assess selection factors. Using the proposed methodology, we also conducted an exploratory analysis that highlights the relative advantages and disadvantages of most of the world's countries as U.S. partners for stability operations.

Currently, various organizations within DoD, including OSD, the COCOMs, and the services, establish foreign partnership priorities based on generally high-level criteria whose application to specific countries is neither completely clear nor consistent nor grounded in empirical analysis. In addition, partnership priorities stem from a number of political and military aims, most of which do not directly relate to building partner capacity for stability operations. Thus the goal of this chapter is to present an objective, transparent, and broadly conceived prioritization method that sets the stage for a detailed, country-by-country examination of potential BPC for stability operations partnerships.

Our exploratory analysis focused on identifying three types of potential stability operations partners. These partner types are defined as follows.

- **Coalition partner.** A willing provider of significant stability operations-related capability in support of coalition operations outside the nation's own borders. A preferred partner demonstrates a

moderate level of *internal* stability, international legitimacy, and strategic affinity with the United States.
- **Regional leader.** An actual or potential provider of capability and leadership for regionally based stability operations that are compatible with U.S. interests. Core regional partners demonstrate a moderate level of internal stability, international legitimacy, and strategic affinity with the United States.
- **Indigenous partner.** A fragile state, preferably receptive to U.S. government assistance and advice, whose deterioration or collapse could pose a significant threat to U.S. interests.

Based on those three partner types, this chapter describes the two partner-selection models—coalition/regional and indigenous—that we used for the exploratory analysis of potential BPC for stability operations partners. Each model is described in terms of its overall structure, as well as its individual attributes, indicators, and scoring methodology.

Country Sample

Most of the world's countries were included in our partner-selection analysis, with a couple of exceptions. First, countries with a population of less than 500,000 (according to World Bank 2005 data) were omitted from our modeling effort. They were considered unlikely to be viable U.S. coalition partners, and any indigenous crisis could be handled by a relatively small number of external troops. Furthermore, data on many of these very small countries did not exist for the measures that were chosen.[1] Second, technically advanced countries (many of which are traditional U.S. allies) were excluded from most (but not all) of our exploratory analysis. We assumed that these countries (see Table 5.1) were already cooperating with the United States on a range

---

[1] The following small countries were omitted from the partner-selection models: Andorra; Antigua and Barbuda; the Bahamas; Barbados; Belize; Brunei Darussalam; Dominica; Grenada; Iceland; Kiribati; Liechtenstein; Luxembourg; Maldives; Malta; Marshall Islands; Federated States of Micronesia; Monaco; Nauru; Palau; Samoa; San Marino; Sao Tome and Principe; Seychelles; Solomon Islands; St. Kitts and Nevis; St. Lucia; St. Vincent and the Grenadines; Suriname; Tonga; Tuvalu; Vanuatu.

Table 5.1
List of Advanced/Major Allied Countries

Australia	Italy
Austria	Japan
Belgium	Korea, Republic of
Canada	Netherlands
Denmark	New Zealand
Finland	Norway
France	Portugal
Germany	Spain
Greece	Sweden
Ireland	Switzerland
Israel	United Kingdom

of security issues and/or were not in need of the type of assistance that is the subject of this study.

Coalition/Regional Model[2]

We decided to construct a single model to prioritize countries for stability operations coalition membership and regional stability operations leadership because of the similarity in the attributes of the two types of partners. These attributes are defined as:

- **Capability.** The potential effectiveness of a state's military contribution to a stability operations coalition.
- **Willingness.** The likelihood that a state will agree to participate in coalition operations outside its borders.
- **Appropriateness.** The net political benefits of a state's participation in a stability operations coalition from a U.S. perspective.

---

[2] See Appendix E for a more detailed technical description of the coalition/regional model.

The structure of the coalition/regional model is shown in Figure 5.1. The overall score for each potential partner country is derived from three attribute scores that measure the country's capability, willingness, and appropriateness. The overall score and the scores for the three attributes are between zero and one [0, 1]. In the base case the attributes are weighted equally (1/3 each) and then summed to give the overall score.

As shown in Figure 5.1, we chose eight quantitative indicators to calculate scores for the coalition/regional model attributes. Similarly, the attribute scores are derived from various indicator scores that are also [0, 1]. Again, in the base case, indictors are weighted equally to determine the attribute score. Indicator scores are derived from data from a variety of authoritative sources, as explained below.

In general, we selected indicators based on their relevance to the attribute under investigation, as well as the availability, comprehensiveness, and credibility of supporting quantitative information. We were not always able to fully satisfy these criteria. For example, military

Figure 5.1
Schematic of the Coalition and Regional Model

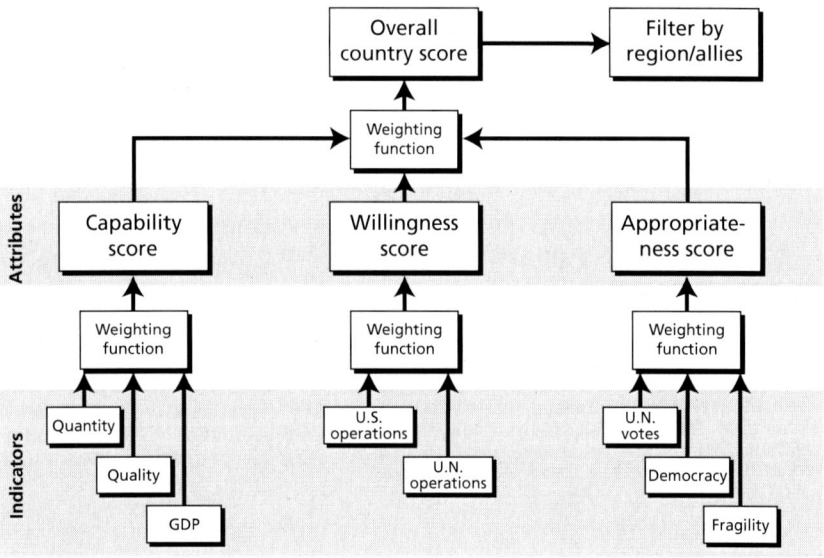

RAND MG942-5.1

quality is dependent on many factors for which comprehensive quantitative data are not readily available. In this case, we had to settle for a reasonable proxy indicator: i.e., military spending divided by military personnel.

We realized that government policy analysts might want to modify the weightings of certain attributes and indicators to reflect the preferences of particular decisionmakers or to demonstrate to DoD leaders how partner country priorities could change as a result of shifting strategic emphases. Thus the indicator and attribute weighting functions of the model can be easily changed so long as the percentile weights given to each indicator and attribute subelement add to one.

Capability Attribute

The capability attribute is derived from three indicators: troop quantity, troop quality, and gross domestic product (GDP).

Troop quantity consists of the number of personnel in a country's armed forces, excluding paramilitary and reserve forces. Data for this indicator were mostly obtained from the International Institute for Strategic Studies, although *Jane's World Armies* and the U.S. State Department web site were consulted for missing data points.

Troop quantity is a measure of a country's ability to contribute to coalition operations. The assumption is that the more troops a country has within its defense establishment, the more troops it can deploy outside its borders. While this is generally true, countries that are already engaged in security activities may be unable to participate in additional operations. For example, Sri Lanka receives a high score in this measure.

The troop quantity score is determined by linearly mapping the number of troops to the score [0, 1]. Any country with 200,000 or more troops gets a score of one. This force level was deemed a sufficient partner contribution for any stability operations-related coalition mission. Although a few countries have higher troop counts, setting the threshold at the maximum possible level (China's 2.25 million troops) would have artificially deflated the capability scores of other potential partners.

Troop quality is extremely difficult to measure in a comprehensive and consistent way. We developed a proxy indicator: military spending per troop, using information gathered primarily from the International Institute for Strategic Studies and, secondarily, from *Jane's* and Global Security.org. This seems an appropriate surrogate for quality in that many countries devote a large percentage of defense budgets to training troops and paying their salaries. However, we acknowledge that not all military training and personnel are relevant to stability operations. Furthermore, a few countries spend a disproportionate share of their defense budgets on capital acquisition for systems that may not aid stability operations. Considering the relatively low training requirements for stability operations, the high threshold for troop quality was set at $20,000 per soldier.

Gross domestic product was included as an indicator of a country's capability to sustain the military and nonmilitary aspects of an external stability operation. For the most part, we relied on World Bank GDP figures from 2005, the most recent dataset that was nearly complete. The CIA's *World Fact Book* provided additional data. The high threshold for the GDP indicator was set at $100 billion, reflecting our view that a large number of countries probably have the economic capacity to support stability operations.

A summary of capability indicator definitions and sources of data is found in Table 5.2.

Willingness Attribute

The willingness attribute is composed of the average contributions to recent U.S.-led operations and U.N.-led operations as a percentage of a country's overall force size. Data on country support for U.S.-led operations were taken from a recently published RAND Arroyo Center report.[3] U.N. deployments were drawn from the United Nations Department of Peacekeeping Operations (DPKO) web site. The latter indicator captures a broad willingness to engage in stability operations that may be especially meaningful for regional partners.

---

[3] Jennifer D.P. Moroney et al., *Building Partner Capabilities for Coalition Operations,* Santa Monica, CA: RAND Corporation, MG-635-A, 2007, pp. 89–95.

Table 5.2
Coalition/Regional Model: Capability Indicators

Indicator	Definition	Source	Details
Quantity	Military forces	International Institute for Strategic Studies	Total armed forces, excluding reserve and paramilitary forces. (2007 *Military Balance*, Table 36.)
Quality	Military forces	International Institute for Strategic Studies	Military spending divided by the number of personnel (from quantity). (2007 *Military Balance*, Table 36.)
Gross domestic product	GDP (US$)	World Bank	2005 World Bank data was used for this report.*

* For more recent GDP data, See World Bank, "Data and Statistics: World Development Indicators 2009." As of February 2010: http://go.worldbank.org/0ROQCBCZG0

The fact that willingness, in the case of both indicators, is defined as a percentage of the overall force deployed gives an advantage to small nations. It is easier for a small nation, such as Fiji, to deploy a large percentage of its armed forces than it is for larger countries in Western Europe or South Asia. An alternative would to be to simply focus on the total number of troops deployed in out-of-country operations. However, this would favor large countries and confound the willingness variable with the capability variable.

Thresholds for deployment indicators were initially set to provide a reasonable spread of willingness scores and to avoid a bias in favor of countries with large armed forces. The high threshold was 0.5 percent of total military forces for contributions to U.S.-led coalition operations and 1 percent of the total force for contributions to U.N.-led peacekeeping operations.

Contributions to Recent U.S.-Led Operations. The aforementioned RAND report (Moroney et al., 2007) examines foreign partner contributions to eight U.S.-led coalition operations: Operation Iraqi Freedom, Operation Enduring Freedom, International Security Assistance Force in Afghanistan, Bosnia, Kosovo, Haiti, Sinai, and Somalia. Our willingness analysis did not include contributions that did not involve military forces, such as providing overflight rights or temporary bases, because this type of assistance is not quantifiable in terms of troop numbers.

Our data source for foreign participation in U.S.-led coalition operations is not entirely suitable as the basis for a willingness measure. First, the study only covers recent operations, some of which do not have a large stability operations component. Second, these operations are not distributed evenly around the globe. A country is more likely to be a willing stability operations participant if its national interests are at stake. For most countries, these interests are highly correlated with the proximity of the conflict to their own territory.

The underrepresentation of U.S.-led operations in Asia and South America in this dataset may introduce a bias into our willingness measure against countries in these regions. However, such a bias may be justified if we assume that future operations will take place in the same regions where they have occurred in the recent past.

Contributions to Recent U.N.-Led Operations. U.N. coalition information was collected from the U.N. DPKO web site. This source includes only U.N. operations directed and supported by the DPKO.[4] We used the total numbers deployed to calculate the average number of troops deployed at any time. We then converted this figure to a percentage of troops deployed, borrowing the total troop numbers from the capability attribute.

Like the U.S. data, the U.N. deployment information covers a relatively short time period and may be unduly influenced by current events. For example, during period we assessed, the United Kingdom was heavily involved in Iraq and Afghanistan which, arguably, may have prevented it from deploying to U.N. operations. Since the United Kingdom exceeds the maximum threshold for U.S.-led operations by a factor of four, our default scoring system "penalizes" the United Kingdom in terms of willingness by weighing contributions to U.N.-led and U.S.-led operations equally. Additionally, some multinational peacekeeping operations do not fall under the mandate of the DPKO. For example, the Australian-led operation in East Timor is not included in our dataset, even though it is sanctioned by the U.N.

---

[4] The data were sampled in six-monthly intervals between October 2003 and April 2007.

Neither willingness indicator takes into consideration the motives of the countries participating in U.S.-led and/or U.N.-led operations. Although some countries may provide troops out of a sense of duty or loyalty, other countries may be primarily motivated by economic incentives. This would seem to be the case for several relatively poor countries such as Fiji, which has the highest score for willingness of any nation in our sample.

Even if data were available to distinguish among various motives, we did not believe this information should be a major element in the willingness measure. There may be cases in which the United States would be disinclined to fund the participation of a country in a coalition operation despite its apparent readiness to deploy troops abroad. However, the rationale for such a decision would probably be captured in the appropriateness or capability attributes of our coalition/regional model.

Some countries receive very low willingness scores despite being allies of the United States. With a willingness score of zero, Israel is the prime example. Our model cannot explain whether this result reflects a true lack of willingness on Israel's part to participate in coalition operations or whether the United States and the U.N. have refrained from requesting Israeli participation for political reasons. This is an example in which our quantitative, macro-level analysis should be supplemented by qualitative, country-specific information. A summary of willingness indicator definitions and sources of data is found in Table 5.3.

Appropriateness Attribute

Appropriateness was included as a litmus test for potential coalition partners because certain capable and willing countries may not meet the grade for political and/or strategic reasons. Furthermore, some less-capable or previously unwilling countries may make acceptable BPC candidates given their strategic alignment with the United States. Appropriateness is a composite measure of democratization, U.N. voting, and fragility. Respectively, they indicate whether a country is politically and/or ideologically similar to the United States, has a similar international outlook, and is domestically stable.

Table 5.3
Coalition/Regional Model: Willingness Indicators

Indicator	Definition	Source	Details
U.S. deployments	Frequency of country assistance to U.S.-led operations	Moroney et al. (2007)	Contributions by nations to eight U.S.-led operations.* The average troop deployment is divided by the force size.
U.N. deployments	Average percentage of total troops deployed to U.N.-led operations	U.N.: Department of Peacekeeping Operations web site	Data was obtained in six-monthly intervals from October 2003 to April 2007 for the number of troops deployed by each country.** This was then averaged and the percentage of the total troops (using the 2005 World Bank data) calculated.

* Moroney et al. (2007), Tables D.1 and D.2, pp. 89–95.
** DPKO web site: http://www.un.org/en/peacekeeping/contributors/

The **democratization** indicator is derived from an Economist Intelligence Unit (EIU) index. This democracy index has five subcomponents: electoral process and pluralism; civil liberties; government functioning; political participation; and political culture. Composite scores based on these subcomponents are used to rank countries and group them into the following categories: full democracies, flawed democracies, hybrid regimes, and authoritarian regimes. The index provides a snapshot of the state of democracy in 165 countries and two territories.

The strength of the democratization indicator is that it is based on a broader concept of democracy than simply holding fair elections; it considers the social and cultural underpinnings of democratic development. The main disadvantage of this indicator is that EIU index results differ from those produced by other democratization indices. In most cases, these differences are slight, but they can be significant. However, this disparity is not unexpected given the difficulty of objectively measuring such a complex social phenomenon.

The **U.N. voting** indicator attempts to capture the degree of strategic affinity that exists between foreign countries and the United States.

As tabulated by the U.S. State Department, this measure includes individual country votes in the U.N. General Assembly and excludes consensus votes. We chose to rely on a country's overall voting behavior rather than its record on DOS-designated "important votes," which was constructed too narrowly for our purposes, focusing on a small number of politically charged issues, many involving Israel and the Palestinians.

U.N. votes are not a perfect measure of policy agreement with the United States. This is particularly evident as the overall level of support for the U.S. voting position decreases. Even some Western European countries score surprisingly low in terms of the percentage of their U.N. votes that align with those of the United States. As a result, the high threshold for this measure was set at 40 percent.

Fragility was measured using the Failed States Index compiled by the Fund for Peace. Using the Conflict Assessment System Tool, this index ranks 177 states according to twelve social, political, and economic indicators. Ratings reflect a state's vulnerability to collapse or conflict, to include: loss of control over its territory, loss of its monopoly on the legitimate use of force, an erosion of its authority to make collective decisions, an inability to provide essential public services, and an inability to interact with other states as a full member of the international community.

A summary of appropriateness indicator definitions and sources of data is found in Table 5.4.

Regional Analysis

To make it easier to examine the rankings of potential partners by region, each country in the coalition/regional model was tagged with the COCOM to which it belonged. For analytical purposes, we assumed a fully operational AFRICOM. This DoD-oriented regional categorization system has pluses and minuses. COCOM areas of responsibility have the advantage of being well defined and of direct relevance when it comes to implementing partner-selection recommendations. In addition, the COCOMs are reasonably well aligned with many other regional frameworks.

Table 5.4
Coalition/Regional Model: Appropriateness Indicators

Indicator	Definition	Source	Details
U.N. votes	U.N. voting (2006)	U.S. State Department	The percentage of times a nation voted with the United States in U.N. general assembly in 2006. Consensus votes are excluded from consideration, as are abstentions.*
Fragility	Failed States Index (2007)	Fund for Peace	A composite measure of national fragility (2007 data).**
Democrati- zation	Democracy Index	Economist Intelligence Unit	A broad measure of democratization by country.***

* http://www.state.gov/documents/organization/82643.pdf, p. 186.

** http://www.fundforpeace.org/web/index.php?option=com_content&task= view&id=229&Itemid=366

*** http://www.economist.com/media/pdf/DEMOCRACY_INDEX_2007_v3.pdf

All categorization schemes that place each country in only one region suffer from similar seam issues. This was addressed in our study by exploring the rankings of certain major countries (those that fall in one COCOM area of responsibility but are of high importance to others) in more than one regional context. These countries are shown in Table 5.5.

Table 5.5
Countries in a Seam Between COCOMs

Country	Actual COCOM	Explored COCOM
Turkey	EUCOM	CENTCOM
Egypt	CENTCOM	AFRICOM
Mexico	NORTHCOM	SOUTHCOM
India	PACOM	CENTCOM

Coalition Analysis

One of the most striking results of our exploratory analysis using the coalition and regional model is the dearth of well-rounded BPC candidates for coalition stability operations. Twenty-two countries, only 14 percent of our sample, had a relatively high overall score of 0.67 or more (on our scale of 0–1.00) in terms of their coalition partnership potential. Of these, 18 are advanced industrial states (many of which are major U.S. allies) that do not require significant capacity-building assistance. By contrast, nearly 75 percent of the countries in our sample had overall scores lower than 0.5 on the coalition partner scale, and 73 countries—45 percent of the total—receive poor ratings overall in our coalition/regional model. None of these countries are advanced industrial states.

Our overall analysis of potential coalition partners suggests that U.S. opportunities for building partner capacity lie largely with states that exhibit considerable stability operations-related shortcomings. We have thus decided to focus our coalition analysis on this problematic group and to devote less attention to the smaller number of advanced states and major U.S. allies that either have a stability operations capacity already or can develop that capacity on their own (as listed above in Table 5.1). Inevitably, our list of advanced/major ally countries is somewhat arbitrary. For example, we have chosen to include new Eastern European NATO allies—as well as one old NATO ally, Turkey—in our exploratory analysis because of the perceived BPC for stability operations benefits that U.S. assistance might bring to these countries.

Analysis of Willingness-Appropriateness Attributes

Our examination of particular coalition/regional attributes—capability, willingness, and appropriateness—begins by considering the full range of potential BPC partners, including advanced/major allied countries. In our analysis of willingness, 76 countries (47 percent of the total) are in the lowest willingness decile. Indeed, 41 countries (about 25 percent of our sample) have a willingness score of zero, thus displaying no inclination to become involved in stability operations. This suggests that finding willing stability operations partners beyond our proven allies will be challenging. Moreover, it makes little sense to try to provide unwilling countries the wherewithal to engage in coalition operations unless the United States can influence their proclivity to participate. Increasing the difficulty of the situation, 77 countries (47 percent) score below 0.5 for both willingness and appropriateness, meaning nearly half of the potential candidates for stability operations BPC appear to be both unwilling and inappropriate.

The primary effect of excluding advanced/major ally countries from the mix of BPC for stability operations candidates is a marked drop in the number of potential coalition/regional partners that are both highly appropriate and highly willing. After excluding the advanced/major ally countries, only ten countries score above 0.5 for both willingness and appropriateness:

- Argentina
- Czech Republic
- Fiji
- Georgia
- Hungary
- Mongolia
- Poland
- Slovak Republic
- South Africa
- Uruguay.

What is striking about the above ten states is that four of them are new NATO allies, and the majority of them have small, relatively weak militaries. Six of them have militaries of 40,000 or fewer personnel. Poland is the only one of these highly willing and appropriate countries with a military that includes more than 100,000 personnel. Of course, the absolute size of a country's military is not necessarily the only, or even most important, measure of an effective regional leader.[5] And countries with mid-sized militaries, such as South Africa and

---

[5] For example, Australia led the multinational effort (INTERFET) to stabilize Timor-Leste after it became independent from Indonesia. This operation required some 9,400 soldiers, of which 4,500 were from Australia and the rest from 19 other countries.

Argentina, may be able to play a leadership role in certain regional contingencies. Furthermore, niche capabilities provided by a number of small countries may make them useful stability operations contributors. However, larger and lengthier contingencies will require countries with the stamina and strength usually associated with large militaries.

Analysis of Willingness-Capability Attributes

The coalition stability operations picture improves if we focus only on the capability scores of countries that are not in the advanced/ major ally group: 28 countries have high capability scores, a significant increase over the three countries that had high overall scores. However, combining willingness and capability significantly decreases the number of attractive partners. Seventy-eight countries (48 percent) have both willingness and capability scores below 0.5. Interestingly, the list of countries with the highest capability and the lowest willingness includes China, Indonesia, Russia, and Turkey.[6] All are significant regional powers and potential global actors that could play a major role, although not necessarily a desirable one in all cases from a U.S. perspective, in future stability operations.

Shifting the Weighting of Willingness Indicators

The willingness score used in the coalition/regional model is a composite of two indicators: participation in recent U.S.-led stability operations, and participation in U.N.-led peacekeeping operations. When these two indicators are weighed equally (as they have been to this point in the exploratory analysis), the model is biased against countries that have contributed only to U.N. operations for a couple of reasons. First, while participation in U.S. operations is an indicator of a country's willingness to work with the United States, and possibly of strategic affinity with the United States, the vast majority of stability/peace-

---

[6] It should be noted that Turkey, Russia, and Indonesia have participated in internationally sanctioned stability operations. Russia, for example, committed 1,200 soldiers to IFOR/ SFOR and 3,200 soldiers to KFOR. However, the contributions of these three countries have been small relative to the size of their militaries. Thus their willingness score is lower than some countries with smaller militaries that have contributed a greater percentage of their forces to recent coalition operations.

keeping operations in the last several decades have been conducted under U.N. auspices. Second, the relatively few recent U.S. stability operations (eight in our database) have had a narrow geographic focus: four in Southwest Asia, two in the Balkans, one in the Horn of Africa, and one in the Caribbean. Perhaps not surprisingly, relatively few U.S. troops have participated in U.N. peacekeeping missions since 1988, with almost none since 2004. In contrast to the lack of U.S. participation, 33 countries were contributing more than 500 police and soldiers to such missions in December 2006 (see Table 5.6). Interestingly, 22 countries are contributing 1,000 or more personnel, and only five of these countries are advanced states/major allies.

The phenomenon of large-scale participation in U.N.-led peace-keeping missions suggests that it would be useful to explore the willingness attribute in terms of this one indicator. Of course, this assumes that the United States would consider BPC for stability operations support to countries that have not as yet contributed to U.S.-led operations.

Table 5.6
Major Military and Police Contributors to U.N. Peacekeeping Operations (in December 2006)

More than 5,000 military and police		More than 500 military and police
Bangladesh		Argentina
India		Chile
Pakistan		Egypt
		Ireland
		Namibia
More than 1,000 military and police		Niger
		Philippines
Benin	Kenya	Poland
Brazil	Morocco	Tunisia
China	Nepal	Turkey
Ethiopia	Nigeria	Ukraine
France	Senegal	
Germany	South Africa	
Ghana	**Spain**	
Indonesia	Sri Lanka	
Italy	Uruguay	
Jordan		

SOURCE: United Nations Department of Peacekeeping Operations, http://www.un.org/Depts/dpko/dpko/contributors/2006/dec06_2.pdf (as of 5 December 2007).

NOTE: Countries in **bold** are advanced/major allied countries.

That said, if involvement in U.N. peacekeeping becomes the proxy for willingness, the result is a large increase in the number of countries in the most willing category, from one country to 27 willing countries. This increase is somewhat offset by the relative weakness and unsuitability of most of these countries. Of those 27 countries now deemed willing, 17 of them score below 0.5 for both capability and appropriateness. Only three states—Argentina, the Slovak Republic, and South Africa—combine high levels of capability and appropriateness with a very high level of willingness to participate in U.N. peacekeeping.

Analysis of Countries with High Capability

One approach to picking BPC partners for stability operations would be to focus on states with high capability, assuming that they could make relatively greater contributions to stability operations and, in some cases, even lead them. In our analysis of this approach, although 28 countries (20 percent of the total) fall into the high-capability category (capability scores above 0.67), only 5 countries (3 percent) can be considered "preferred," meaning they score above 0.5 in both willingness and appropriateness. These countries are:

- Argentina
- Czech Republic
- Hungary
- Poland
- South Africa.

Based on this analysis, it is clear that most high-capability countries are generally unwilling to participate in stability operations and/or are inappropriate for such operations.

Regional Analysis

A regional/coalition model analysis of the countries in the three geographic combatant commands where the Army has the greatest BPC for stability operations responsibilities—CENTCOM, AFRICOM, and SOUTHCOM—generally confirms the results of the previous attribute analysis. However, this investigation also points to a few countries that exhibit the characteristics of potential regional leaders of stability operations.

U.S. Central Command

Although of great strategic importance to the United States, CENT-COM is not a particularly promising region for stability operations partnerships. Of the 20 countries in this grouping, all but one have overall scores below 0.5.

Willingness and appropriateness scores are particularly grim. Despite the extensive involvement of the United States in the CENT-COM region, few countries appear willing to become involved in stability operations. With a willingness score in the 0.6 decile, one relatively bright spot is Pakistan. Moreover, if only contributions to U.N. peacekeeping operations are used to calculate willingness, Pakistan's score shifts to 0.9 due to the many peacekeeping missions it has supported. As of August 2007, Pakistan was the top contributor to U.N. peacekeeping missions, with forces in the Democratic Republic of Congo (3,850), Liberia (3,401), Sudan (1,570), and the Ivory Coast (1,123). In addition, in 1994 Pakistan was the largest contributor of troops to UNOSOM II in Somalia (7,000 soldiers or more than 37 percent of deployed forces).

Our analysis also indicates that most countries in CENTCOM have modest military capabilities—only five had capabilities that could be considered "high" (above 0.67): Egypt, Iran, Pakistan, Saudi Arabia, and the United Arab Emirates. Of those, only Pakistan has shown a willingness to participate in stability operations.

U.S. Africa Command

Despite the region's need for well-trained and equipped peacekeepers, Africa has very few strong BPC for stability operations candidates.[7] Our

---

[7]   According to the commander of the proposed U.N.-African Union Darfur force, few African countries have enough soldiers that meet U.N. peacekeeping standards and most can not sustain in the field for six months. Opheera McDoom, "Ethiopia Pledges 5,000 Peacekeepers to Darfur," *Reuters*, 4 October 2007.

analysis found only one state—South Africa—that has an overall score above 0.6. Furthermore, only four states—Algeria, Morocco, Nigeria, and South Africa—have relatively high capability scores. Indeed, some 65 percent of AFRICOM countries have capability scores of 0.3 or below.

Nevertheless, Africa has a tradition of involvement in stability operations in support of the U.N. and regional organizations. When only support for U.N. operations (and not U.S.-led coalition operations) is considered, there is a significant positive shift in the willingness scores. This is likely due to two factors: the United States has conducted only one recent stability operation in Africa (Operation Restore Hope in Somalia), and many of the region's countries lack the capability and interest to engage in out-of-area operations.[8]

Algeria and South Africa are the most capable countries in AFRICOM. South Africa receives the region's highest overall score because of its willingness to participate in stability operations and its relatively high appropriateness rating.[9] Although even more capable than South Africa, Algeria has a low overall score because of its poor appropriateness score and lack of involvement in stability operations.

U.S. Southern Command

Despite the region's historical ambivalence toward the United States, Latin America and the Caribbean provide some BPC for stability operations opportunities. Although most countries in the region have low overall scores, three stand out as potential stability operations partners and/or regional leaders: Argentina and Chile, with overall scores in the 0.6 decile, and Brazil, with a score in the 0.5 decile.

An interesting characteristic of the SOUTHCOM region is its relatively high level of appropriateness. Twelve countries (54 percent)

---

[8]   Morocco contributed soldiers in support of U.S. operations in both Bosnia and Somalia, although in Bosnia, Morocco's contribution predated NATO's commitment of forces and was part of the French multinational division. Morocco's contribution to UNITAF was substantial, amounting to some 1,300 soldiers. Zimbabwe contributed some 1,000 soldiers to Operation Restore Hope in Somalia.

[9]   Only Mauritius has a higher appropriateness score than does South Africa. However, Mauritius is a militarily weak country that has not participated in stability operations.

score above 0.5 for appropriateness—well above the global average of 29 percent. Worldwide, more than 50 percent of the countries are clustered in the bottom four deciles of the appropriateness scale. By contrast, only 13 percent of SOUTHCOM countries are in this position. Driving this phenomenon are two factors: the region's high level of democratization and its relatively high strategic affinity with the United States.

The apparent willingness of some SOUTHCOM countries to participate in stability operations increases if we focus on participation in U.N. operations rather than on U.S.-led coalition operations.[10] Argentina, Brazil, Chile, and Uruguay are all important contributors to the United Nations Stabilization Mission in Haiti. In addition, Uruguay and Chile have also deployed significant forces in support of U.N. operations in Congo and Cyprus, respectively.

Indigenous Model[11]

The RAND team developed the indigenous model as a means for the Army (and others) to assess fragile states, preferably those that would be receptive to U.S. government assistance and advice, whose deterioration or collapse could pose a significant threat to U.S. interests. The calculation of the overall partner nation scores in the indigenous model is similar to, but simpler than, that used in the regional/coalition model. In the indigenous model, only two major attributes are considered: fragility and receptivity. These attributes are defined as follows:

- **Fragility:** a potential partner regime's vulnerability to collapse
- **Receptivity:** a state's openness to U.S.-led and/or U.N.-led deployments within its borders, combined with a minimal synergy with U.S. national security objectives.

---

[10] Colombia, the Dominican Republic, El Salvador, Honduras, Nicaragua, and Uruguay have all made small contributions to U.S.-led operations in Iraq and the Sinai. Chile contributed to recent brief U.S.-led operation in Haiti.

[11] See Appendix E for a technical description of the indigenous model.

With only two major attributes and three indicators, the indigenous model is similar to, but simpler than, the coalition/regional model (see Figure 5.2). The fragility attribute has just one indicator, which is derived from the Failed States Index. The receptivity attribute has two indicators: U.N. votes and the EIU's democratization index. These measures are also used in the coalition/regional model, although the rationale for selecting them is different.

The default weightings used to determine the overall score in the indigenous model are 2/3 for fragility and 1/3 for receptivity. This varies from the uniform weights used elsewhere in both models. This reflects the idea that the purpose of this model is to find fragile countries that are also receptive. Of course, analysts have the option to change this weighting function.

Structurally, the indigenous model differs from the coalition model in its use of five U.S. strategic interests as post-processing filters, which come into play after countries have been ranked according to their weighted attributes. These include:

Figure 5.2
Schematic of the Indigenous Model

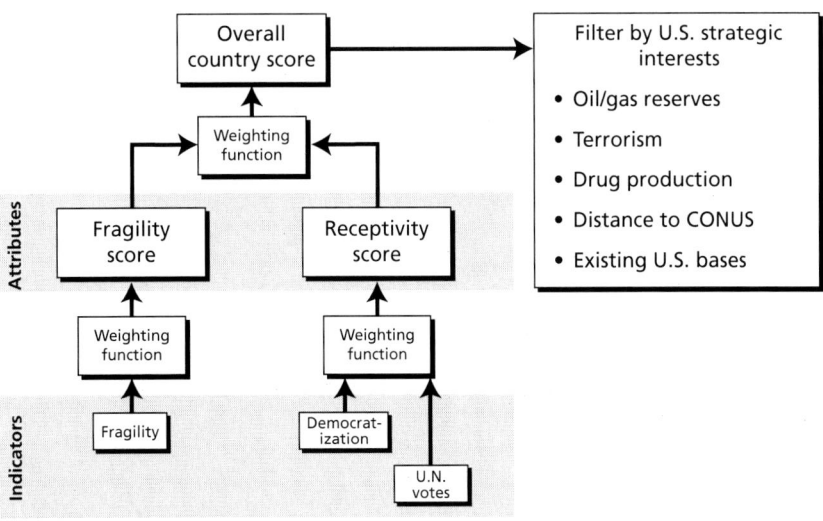

- Maintaining access to energy sources
- Countering international terrorism
- Countering illicit drug production
- Protecting the near abroad
- Protecting overseas bases.

This model allows analysts to sort countries by particular strategic interest or to find countries in which the United States has any strategic interest.

Fragility

As in the regional/coalition model, the indigenous model's fragility attribute is also based on the Fund for Peace's Failed States Index, which identifies four levels of fragility: alert, warning, moderate, and sustainable. However, the indigenous model diverges from the Failed States Index in one way. In the Failed States Index, the moderate level of fragility is associated with countries on par with the United States, which we do not consider realistic candidates for indigenous BPC support. Therefore, we have conflated the "moderate" and "sustainable" categories into a single category.

Democratization

The democratization indicator is derived from an Economist Intelligence Unit index, which has five subcomponents: electoral process and pluralism; civil liberties; government functioning; political participation; and political culture. Composite scores based on these subcomponents are used to rank countries and group them into the following categories: full democracies, flawed democracies, hybrid regimes, and authoritarian regimes. The index provides a snapshot of the state of democracy in 165 countries and two territories.

Democratization was selected an indicator of receptivity based on the assumption that democratic governments are more willing and able to effect reform. We acknowledge that this may not always be the case. In some instances, an authoritarian government, with a strong interest in maintaining good relations with the United States, may be more capable of executing significant political changes than a fractious democratic regime, especially if these changes do not meet with the approval of important domestic constituencies. However, democratic governments have a better record of maintaining a course of reform over the long term without provoking widespread social disruption.

United Nations Votes

The U.N. votes indicator reflects a country's receptivity to U.S. foreign policy goals. Our assumption is that the more willing a country is to vote with the United States in the U.N., the more likely it is to accept U.S. aid and use it in a way that conforms to U.S. preferences. Although this assumption may not be valid at all times for every country, a country's U.N. voting record tends to highlight countries with regimes that are actively hostile to the United States and thus would not be trustworthy recipients of American assistance. As in the coalition model, the high threshold for this indicator in the indigenous model was set at 40 percent, which roughly corresponds to the level of voting alignment between the United States and its traditional European allies.

U.S. Strategic Interests

Because of the large number of countries in need of security assistance, it is sensible to develop a mechanism for segregating those in which the United States has compelling strategic interests. Although we do not claim that our model fully incorporates these interests (some of which are politically contentious and/or difficult to quantify), we have chosen five that approximate the range of interests that are important to most U.S. policymakers.

Given the uncertain fidelity of our data, countries were assessed as either being of strategic interest to the United States in a particular area (in which case they received a score of 1) or not (in which case they received a score of 0). No intermediate scores were calculated. Additionally, strategic interests were not added together or combined in any way. Our summary of U.S. strategic interests indicates only whether or not a country met at least one of the criteria.

The following sections provide more information on the five strategic interests that we selected for our indigenous model.

Oil/Natural Gas. Energy security is a key concern for the United States and is an important component of the U.S. interest in the Middle East, Africa, and Central Asia. To assess the extent of this particular interest, we used a list of countries' proven oil and natural gas reserves compiled by the U.S. Department of Energy. This list contains three estimates for each country. So as not to unduly restrict our grouping of strategic countries, we used the largest of these estimates except when there was one estimate that deviated wildly from the other two. We then calculated the average price over a two-year period for oil and gas, and combined the results in dollar terms.

After examining various options, we chose to set the minimum oil/gas threshold for a strategic country at 1 percent of the total production value of the countries in the indigenous model. According to this definition, the United States has a strategic energy interest in the countries shown in Table 5.7.

Terrorism. The United States has a general interest in reducing the incidence of terrorism around the world. That said, the country's most pressing interest is in preventing terrorism directed at U.S. citizens. For this reason, we chose to use as our terrorism indicator countries in which terrorist attacks have been directed against U.S. civilians and/or U.S. property. An alternative indicator might have been the national origin of the perpetrators of terrorist strikes against Americans. However, this information was not available in the large majority of cases.

Table 5.7
Countries with Significant Energy Reserves That Are of Strategic Interest to the United States

U.S. Strategic Interests: Energy Reserves	
Iraq	Saudi Arabia
Nigeria	United States
Russian Federation	Libya
Venezuela	Kuwait
China	Canada
Iran	Qatar
Kazakhstan	Norway
Algeria	United Arab Emirates

The National Counterterrorism Center database includes 123 incidents in 43 countries that meet our U.S. targeting criteria. The incidents that were catalogued included the following types of terrorist attacks: Islamic extremism, political violence, and environmental terrorism. Establishing a minimum threshold of two attacks reduces the number of countries in which the United States has a strategic interest due to terrorism to the 20 countries shown in Table 5.8.

There are some countries that have been associated with terrorism that are missing from the above list. Iran, Syria, and North Korea are prominent examples. Although these countries have been accused of sponsoring the activities of terrorist groups in other countries, the National Counterterrorism Center database does not show any terrorist attacks directed against U.S. targets that occurred on their soil. In any event, these alleged state sponsors of terrorism are unlikely recipients of U.S. BPC assistance.

Illegal Drugs. The three elements of the drug problem that we could have addressed were production, trafficking, and usage. We concluded that usage was a domestic issue and therefore not relevant to our model. In addition, we considered that our proximity measure covered the countries most implicated in U.S.-oriented drug trafficking. This left drug production as the basis for our strategic interest indicator.

Table 5.8
Countries with Terrorist Attacks Against U.S. Targets

Terror Attack Locations	
Iraq	Philippines
Afghanistan	Serbia
Israel (includes	Egypt[a]
Occupied Territories)[a]	Thailand
Pakistan[a]	India
Bangladesh	Jordan
Nigeria	Greece
Nepal	Saudi Arabia[a]
Turkey	Italy
Indonesia[a]	Argentina

[a] Countries with two or more Islamic extremist attacks.

Although comprehensive data on drug production per country is lacking, the vast majority of the world's coca and opium (the most important raw ingredients of illegal drugs) comes from a few well-known countries. The following countries were identified in the U.N. *2007 World Drug Report*:[12]

- Afghanistan
- Bolivia
- Colombia
- Laos
- Mexico
- Myanmar
- Pakistan
- Peru.

Geographic Proximity. Despite its status as a global power, U.S. interest in certain countries is very much influenced by their proximity to the U.S. homeland. A classic example would be Cuba, 90 miles off the Florida coast, whose government allowed the deployment of Soviet nuclear weapons during the early 1960s, provoking the most serious crisis of the Cold War. More recently, geography has enabled other problems to spread to American shores. These problems include the traffic in illegal drugs, mentioned above, as well as uncontrolled migration as a result of economic privation, civil war, political repression, and natural disasters in neighboring countries.

For the purpose of our indigenous analysis, countries located in North America, Central America, and the Caribbean were deemed to be of strategic proximity to the United States, as were South American countries bordering the Caribbean, in particular, Colombia and Venezuela.

U.S. Bases. The final U.S. strategic interest that we examined was countries hosting U.S. military bases. The list of such countries was extracted from a FY 2006 report issued by the Office of the Deputy Under Secretary of Defense (Installations and Environment). For our analysis, we included any country with an active U.S. military instal-

---

[12] United Nations Office on Drugs and Crime, *2007 World Drug Report*.

lation (i.e., provider of employment), with the exclusion of Iraq and Afghanistan, whose operational bases were considered temporary. Selected countries with U.S. bases are shown in Table 5.9.

Indigenous Exploratory Analysis

Our exploration of indigenous stability operations partners yields two basic conclusions. First, although a large number of fragile states could potentially benefit from BPC support, only a small subset of these countries are of strategic importance to the United States, according to our modeling criteria. Second, those states that are most in need of help are often among the least receptive to stability operations-related aid.

Fragility and Receptivity

Most of the world's countries are fragile. Based on our analysis, nearly 80 percent (110 countries) can be found in the bottom five deciles of the model's fragility range.[13] Furthermore, more than 54 percent (75 countries) score below 0.5 for both the fragility and receptivity indicators. The high correlation between high fragility and low receptivity is problematic, as it suggests that those countries most in need of support

Table 5.9
Location of United States Overseas Bases

Countries with U.S. Bases		
Australia	Germany	Oman
Bahrain	Greece	Peru
Belgium	Indonesia	Portugal
Canada	Italy	Qatar
Colombia	Japan	Singapore
Cuba	Kenya	Spain
Denmark	Korea, Republic of	Turkey
Ecuador	Kuwait	United Arab Emirates
Egypt	Netherlands	United Kingdom
France	Norway	

---

[13] This calculation excludes advanced/major allied countries.

are also the least likely to accept it or use it efficiently. In addition, poor receptivity could undermine U.S. domestic support for BPC for stability operations assistance. Governments that are perceived as corrupt, inefficient, and/or embracing antithetical values are unlikely to garner sustained support from the U.S. Congress or the public.

Most Fragile States

Thirty-one countries are considered most fragile and in the "Alert Zone," according to the Fund for Peace's Failed States Index. These fragile states include such large and regionally important countries as Bangladesh, Nigeria, and Pakistan. At the top of the fragility list are Chad, Cote d'Ivoire, Zimbabwe, Iraq, and Sudan. Ninety percent of these most fragile countries (28 out of 31) score below 0.5 for both fragility and receptivity.

Regionally, more than half of the most fragile states and nearly all of the states in the most fragile decile are located in AFRICOM.

Strategic Interests

Focusing on states in which the U.S. maintains two or more strategic interests (according to our definition) could narrow the number of indigenous BPC for stability partners to 16 (see Table 5.10).

Table 5.10
States with Two or More Strategic Interests

Afghanistan	Nigeria
Colombia	Pakistan
Cuba	Peru
Egypt	Qatar
Indonesia	Saudi Arabia
Iraq	Turkey
Kuwait	United Arab Emirates
Mexico	Venezuela

NOTE: Only Colombia is associated with more than two strategic interests. Countries in **bold** are Fund for Peace "Alert Zone" countries.

The five countries with the highest indigenous partner scores are Afghanistan, Colombia, Iraq, Nigeria, and Pakistan. While several strategically important states are quite small (Kuwait and Qatar), most are medium to large nations. Four countries—Indonesia, Mexico, Nigeria, and Pakistan—have populations in excess of 100 million, and Egypt, with a population of some 80 million, is also quite large. Such countries could require significant BPC for stability operations assistance and/or a massive operation to restore stability should they ever collapse.

As one would expect given their strategic importance, many of the countries in Table 5.11 are already receiving substantial security assistance from the United States. Seven are partners with the United States in the volatile and oil-rich Middle East. Colombia is the top recipient of U.S. counterdrug support. Some strategically important countries, such as Cuba and Venezuela, are ineligible for BPC support as long as their current governments remain in place. Of particular interest is Nigeria, which receives comparatively little U.S. security assistance. Nigeria's poor receptivity score highlights the difficulty of providing effective stability operations aid to key strategic states. This point is reinforced by the six U.S. allies in the Middle East that have receptivity scores between 0.2 and 0.3.

Conclusion

As stated in the introduction to this chapter, the Army and DoD currently lack rigorous methods for prioritizing foreign partners, not to mention foreign partners whose capacity for stability operations it wishes to increase. Although the two partnership prioritization models we have presented in this chapter are far from ideal, they provide defense policy analysts a simple, transparent, and empirically based tool that demonstrates to decisionmakers how different operational goals and shifting partner preferences can affect the priority assigned to various partner countries, which in turn can affect the distribution of resources designed to build partner capacity for stability operations. It is hoped that such understanding by DoD policymakers will contribute to the development of a clear and holistic strategy for prioritizing partners in this particular security cooperation area.

For example, in our analysis of potential BPC for stability operations partners, willingness to participate in coalition/regional operations is sensitive to how that particular attribute is defined. There are many countries that have made significant contributions to stability operations that have not participated in U.S.-led coalitions. Other states have not as yet contributed to either U.N. or U.S. operations, but score relatively well in terms of capability and appropriateness. Thus there may be a high payoff if the United States devises a coalition/regional BPC strategy that encourages these countries to cooperate with the United States in stability operations. Alternatively, the United States should consider working through the U.N. or regional security organizations, since some stability operations capable countries are more willing to participate in coalitions that are not led by the United States but may nevertheless support U.S. interests.

Our indigenous analysis indicates that only a small number of the world's many fragile states are of strategic interest to the United States. Furthermore, many countries in need of support may have great difficulty either accepting U.S. aid or using it effectively.

For these reasons, it would be beneficial for the United States to develop a highly selective strategy for building partner capacity for stability operations that is nested within the National Security and National Military Strategy.

CHAPTER SIX

Recommendations

There are a number of principal recommendations based on our study's analyses of BPC and stability operations guidance, baseline activities, programmatic effectiveness, and potential partner countries. Given the ambitious scope of the project, the evolving character of the topic, and constraints on researchers' access to complete datasets, these recommendations should be treated as more suggestive than definitive. Significantly more data collection and analysis will be required before DoD and other government officials have the knowledge and understanding necessary to bring the various aspects of BPC for stability operations policy into alignment.

First, BPC and stability operations guidance needs to be more clearly defined and better integrated so that U.S. government agencies understand not only primary strategic objectives but also their respective roles and missions. Mechanisms for aligning Army, DoD, and national BPC for stability operations strategy, planning, and resourcing should be constructed. Ideally, overall security sector assistance would be jointly managed by the departments of State and Defense. This could result in interagency objectives for employing and developing DoD and DOS resources and capabilities for building partner capacity, as well as standardized procedures for formulating detailed BPC "roadmaps" for priority partners.

BPC and Stability Operations Guidance

In developing their part of a larger BPC for stability operations strategy, DoD and the U.S. Army should seek to understand the extent and effectiveness of the BPC for stability operations activities that they, their interagency partners, and major U.S. allies are currently conducting, and promote BPC for stability operations coordination and cooperation where possible. Specifically, DoD's new emphasis on working "by, with, and through" partners requires DoD to

- Flesh out the list of essential BPC for stability operations capabilities (perhaps along the lines that we have suggested in this report).
- Distinguish between directly providing stability operations assistance and BPC for stability operations as well as between stability operation-related activities and stability operations-useful activities.
- Consider a range of BPC for stability operations contexts (coalition, regional, and indigenous) when selecting and prioritizing potential partners.

Baseline Activities Analysis

At a minimum, our baseline analysis of BPC for stability operations activities (as detailed in Chapter Three) suggests that the military should improve its visibility into security cooperation activities relevant to building partner capacity for stability operations. In particular, the Army should

- Cooperate with the COCOMs to increase the quality and quantity of their TSCMIS data, especially those related to funding and personnel.
- Ensure that its new security cooperation database is flexible enough to be used for analytical as well as operational purposes.

Once it has acquired an overall understanding of ongoing BPC for stability operations activities, the Army should

- Increase the number and extent of its BPC for stability operations activities in certain regions, such as AFRICOM, where its programs are relatively scarce but where arguably the demand is growing.[1]
- Make a concerted effort to learn from the BPC for stability operations experience of its allies, in particular the UK and France, in several key areas such as trainer selection, mode of deployment, training of the trainers, and career implications for the trainer.
- Re-evaluate its methods of delivering stability operations assistance to various partners—e.g., direct U.S. help or BPC aid, dedicated activities linked to stability operations capabilities or general-purpose activities that could serve as building blocks for stability operations and other kinds of operations.

Detailed Activities Assessment

Although the U.S. government currently lacks an assessment framework for BPC for stability operations, our six-step assessment approach and detailed case studies of a range of BPC programs (as detailed in Chapter Four) indicated that the Army Service Component Commands should continue to assist the COCOMs in developing a holistic framework that

- Is planned and resourced over a period of several years, involving all relevant U.S. military and civilian agencies and allies.
- Targets multiple countries throughout a region.
- Employs a variety of security cooperation tools that are packaged and sequenced for each partner country.

---

[1] This recommendation may be difficult to implement in the short run given the requirements for forces in Iraq and Afghanistan. However, building partner capacity is one way of sharing the security burden with other countries so that the United States may avoid having to employ large numbers of U.S. forces for stability operations and other purposes.

- Considers the indigenous requirements of partners, thereby reducing the need for direct U.S. military assistance and increasing the incentive for partners to engage in future coalition operations with the United States.
- Takes into account various external factors that enable or constrain security cooperation activities in different contexts so they can be incorporated into programmatic and regional planning for BPC for stability operations.

Analysis of Potential Partners

In general, our partner analysis using the regional/coalition and indigenous models could support divergent courses of action (as detailed in Chapter Five). The apparent scarcity of high-potential partner nations could

- Justify a narrowing of U.S. government BPC for stability operations efforts to avoid squandering limited security resources.
- Serve as an impetus for greatly increasing the amount of BPC for stability operations resources in an attempt to bring more countries to the point where they might become effective coalition, regional, or indigenous partners.

A more specific recommendation is to place greater focus on coalition and regional candidates that have a demonstrated willingness to participate in U.N. deployments. Defining willingness in terms of past involvement in U.S.-led coalition operations may unnecessarily constrain DoD's choices. However, building relationships with countries that have seldom or never entered into coalitions with the United States will require prolonged and creative engagement and, perhaps, a greater willingness to share control over stability operations than would be the case with long-standing U.S. partners. Ironically, the best indigenous BPC for stability operations partners are probably countries that have relatively lesser need for U.S. assistance. Because few countries are both fragile and receptive, attempts to build indigenous stability operations capacity should, in many cases, be based both on the degree of a country's internal weakness and on the salience of the U.S. strategic interest in that country.

Additional Research

Although we have explored key aspects of a U.S. government BPC for stability operations strategy using several analytical methods, additional research is required to fill some of the identified gaps. First, primarily because of data constraints, we were unable to integrate our baseline activities analysis, detailed activities assessment, and analysis of potential partners to the degree that would enable a comprehensive alignment of suitable BPC for stability operations partners with relevant and effective security cooperation activities for resource management purposes. Second, due to our study sponsor's responsibility for U.S. Army security cooperation, we deliberately chose to focus on pre-conflict and, to a certain extent, post-conflict BPC for stability operations activities. We did not analyze DoD or other U.S. government programs that advise and support partner nations involved in ongoing conflict operations. As a result, our analysis may have inadvertently reinforced DoD's traditional tendency to separate, both conceptually and organizationally, security cooperation from conflict operations. Third, our study did not address the important issue of *building partnership capacity*: that is, U.S. capability requirements for force structure, training, personnel et cetera to execute a broad range of activities in support of a BPC for stability operations strategy. Finally, because the research phase of the study ended in 2007, this report does not include more recent Office of Secretary of Defense (OSD) and Joint Staff guidance on building partner capacity for stability operations and how that affects the Army. Additional research will be needed in order to address these analytical deficiencies.

Defining Capabilities for Stability Operations

This appendix defines the generic capabilities associated with the Department of Defense stability operations major mission element (MME) and the U.S. Army's logical line of operation (LLOs) discussed in Chapter Two. Drawing on official Department of State and DoD guidance, each section lists the overall stability operations MME (and if appropriate, each subordinate mission), followed by working definitions of specific stability operations capabilities.

Establish and Maintain a Safe and Secure Environment

Several subordinate missions flow from this MME.

Disposition of Armed and Other Security Forces, Intelligence Services, and Belligerents

This subordinate mission has three capability requirements.

Conduct Peace Operations

Working definition. The ability to conduct tactical military operations designed to monitor, facilitate, or enforce the implementation of an agreement, either negotiated or imposed, that are intended to create the condition for conflict resolution in order to establish and maintain peace. This includes ceasefires, truces, or other such agreements.[1]

---

[1] Derived from Operational Level (OP) Task 3.3.1, Conduct Peacekeeping Operations in the Joint Operations Area; Task 3.3.2, Conduct Peace Enforcement Operations in the

This capability includes the ability to monitor, supervise and enforce ceasefires and the disengagement of hostile forces; establish and control buffer and demilitarized zones; conduct limited military operations (raids, strikes, show of force) to enforce the agreements; develop and support confidence-building measures between indigenous belligerents; facilitate prisoner-of-war exchanges; and to support and enforce the military, political, and economic terms of the agreement.

Essential Tasks Matrix (ETM) subtasks. Cessation of hostilities, enforcement of peace agreements, and/or other arrangements.

Conduct Disarmament, Demobilization and Reintegration Operations

Working definition. The capability to conduct disarmament, demobilization, and reintegration operations in support of war-to-peace transitions by reducing or eliminating belligerent armed forces and the supply of armed weapons by facilitating the return of former combatants to sustainable civilian livelihoods. This capability includes the ability to establish and enforce weapons control; secure, store, and dispose of weapons; establish and decommission demobilization camps; ensure the adequate health, food provisions, and security of ex-belligerents; gather and disband the structural element of belligerent groups; monitor and verify demobilization; and ensure the safety of quartered personnel and their families. This task also includes the capability to plan, monitor, and execute programs to reintegrate belligerents back into their communities.[2]

ETM subtasks. Disarmament, demobilization, reintegration of combatants.

Develop and Sustain Armed Services and Intelligence Forces

Working definition. The capability to develop and sustain security and intelligence forces that can conduct legitimate self-defense

---

Joint Operations Area. JCS, *Universal Joint Task List*, CJCSM 3500.04D, August 2005, p. B-C-C-69.

[2] Derived from Peace and Security program element 3.2, Disarmament, Demobilization, and Reintegration (and its associated sub-elements), in *Foreign Assistance Standardized Program Structure and Definitions*, 20 October 2006. See also FM 3-07.31, 2003, pp. III-1, IV-28 to IV-30.

operations to maintain control or regain control over national territory. This includes the ability to create professional military and intelligence forces that are transparent and accountable to the civilian government.[3] This capability includes the ability to identify the future roles, missions, and structures of the military and intelligence forces; to vet senior personnel and others for past abuses; to train and equip; to establish transparent entry and promotion procedures; and to promote civilian oversight.

ETM subtasks. Disposition and constitution of national armed services; disposition and constitution of national intelligence service(s).

Territorial Security

Territorial security has three subcomponents that are not easily merged.

Establish and Maintain Border and Boundary Control

Working definition. The ability to secure borders and boundaries and to regulate the movement of people and goods across them.[4]

ETM subtasks. Border and boundary control.

Establish and Maintain Freedom of Movement

Working definition. Capability to ensure the uninhibited flow of civilian traffic and commerce so as to allow the resumption of normal activity and to guarantee the right of transit of nongovernmental organizations, noncombatants, and stability operations personnel. This includes ensuring that lines of communications are free of obstacles and unobstructed by belligerent or other forces.[5]

ETM subtasks. Freedom of movement.

---

[3] Derived from Peace and Security program subelement 3.6.1, Territorial Security and Governing Justly and Democratically program element 2.5, Governance of the Security Sector, in *Foreign Assistance Standardized Program Structure and Definitions*, 20 October 2006.

[4] Derived from the FM 3-07.31 definition of border control. Air Land Sea Application Center, *Multi-Service Tactics, Techniques and Procedures for Conducting Peace Operations*, FM 3-07.31, October 2003, pp. III-1.

[5] Derived from HQDA, *Stability Operations and Support Operations*, FM 3-07, February 2003, pp. 2-2, 4-9 to 4-10, 4-17.

Establish an Identification Regime

Working definition. The capability to plan, establish, and enforce a civilian identification regime, including documents relating to personal identification, property ownership, court records, voter registries, birth certificates, and driving licenses.[6]

ETM subtasks. Identification issues.

Public Order and Safety

Provide Interim Public Order

Working definition. The ability to ensure a lawful and orderly environment and to suppress criminal behavior. This capability includes the ability to protect vulnerable noncombatants and to engage in crowd and disturbance control operations.[7]

ETM subtasks. Protection of noncombatants, interim policing, controlling crowds, and disturbances control.

Explanation. This capability focuses on the ability to provide immediate and emergency public order in the absence of effective and capable civilian police or other law-and-order forces.

Conduct Civilian Police Operations

Working definition. The capability to establish and sustain effective, professional, and accountable law enforcement services with the capacity to protect persons, property, and democratic institutions against criminal and other extralegal elements.[8]

ETM subtasks. Indigenous police personnel, essential police facilities, accountability/oversight.

---

[6] This is a part of population and resource control operations. See HQDA, *Counterinsurgency*, FM 3-24, December 2006, p. 5-21; HQDA, *The Army Universal Task List*, FM 7-15, July 2006, pp. 6-112 to 6-113;

[7] Derived from the definition for Army Tactical Task 7.7.2.2, Provide Law and Order. HQDA, *Army Universal Task List*, FM 7-15, July 2006, p. 7-41.

[8] Derived from the Peace and Security program subelement 3.7, Law Enforcement Reform, Restructuring, and Operations definition in *Foreign Assistance Standardized Program Structure and Definitions*, 20 October 2006.

Explanation. The State Department sectoral task "indigenous police" has been moved from the justice technical sector to the security technical sector because of the key role that police forces play in ensuring security in a stability operations environment.

Conduct Emergency Clearing Operations

Working definition. The ability to conduct clearance operations to remove or neutralize mines and unexploded ordnance that are an immediate threat to civilians and stability operation personnel.[9]

ETM subtasks. Clearance of unexploded ordnance.

Explanation. This unexploded ordnance clearance subtask appeared sufficiently different from the others in this task area to warrant separate treatment.

Protection of Indigenous Individuals, Infrastructure, and Institutions

Provide Protective Services

Working definition. The ability to protect key political and societal leaders from assassination, kidnapping, injury, or embarrassment.[10]

ETM subtasks. Key leaders, witness protection.

Protect Critical Installations and Facilities

Working definition. The ability to protect and secure strategically important infrastructure, facilities, sites, and institutions from hostile actions. This includes the securing and protection of private property and factories, religious sites, cultural sites, military facilities, critical infrastructure and natural resources, and public institutions. It also includes securing facilities containing documents and other evidence related to key ongoing or potential U.S. investigations and prosecutions.[11]

---

[9] Derived from HQDA, *Army Universal Task List*, FM 3-15, July 2006, p. 5-4; HQDA, *Stability Operations and Support Operations*, FM 3-07, February 2003, pp. 2-2, 2-8 to 2-9.

[10] Derived from the definition for Army Tactical Task 5.3.6.1, Provide Protective Services for Selected Individuals. HQDA, *Army Universal Task List,* FM 7-15, July 2006, p. 5-74.

[11] Derived from the definition for Army Tactical Task 5.3.5.5.5, Conduct Critical Installations and Facilities Security. HQDA, *Army Universal Task List,* FM 7-15, July 2006, p. 5-66.

ETM subtasks. Private institutions, critical facilities, military facilities, public institutions, evidence protection.

Explanation. The tasks of protecting individuals and protecting property have been separated.

Protect Reconstruction and Stabilization Personnel and Institutions

Working definition. The capability to provide physical security and logistical support for civilian personnel and facilities engaged in stability operations. This includes convoy security, the physical security of relief supplies, aid workers, and infrastructure such as roads and airfields.[12]

ETM subtasks. Official civilian stabilization and reconstruction personnel and facilities, contractor and nongovernmental organization (NGO) stabilization and reconstruction personnel and facilities.

Security Coordination

Coordinate Indigenous and International Security Forces and Intelligence Support

Working definition. The ability to coordinate between elements of the stability operation (indigenous, international, multinational, NGO) for the purposes of accomplishing the operations objectives.[13] This includes the ability to integrate command, control, and intelligence and information-sharing arrangements between international military, constabulary, and civilian police forces and between the international and indigenous security forces. It also includes the capability to provide integrated intelligence to the international security and police forces and to coordinate military and civilian command, control, and intelligence and information-sharing arrangements.

---

[12] Derived from Peace and Security program subelement 3.1.6, Armed Physical Security, in *Foreign Assistance Standardized Program Structure and Definitions*, 20 October 2006. See also task ST 4.3.2, Provide Supplies and Services for Theater Forces. JCS, *Universal Joint Task List*, CJCSM 3500.04D, August 2005, p. B-C-B-66, and HQDA, *Stability Operations and Support Operations*, FM 3-07, February 2003, p. 4-10.

[13] Derived from the Joint Interagency/international/multinational/ NGO Coordination tier 1 joint capability area. *Joint Capability Areas Tier 1 and Supporting Tier 2 Lexicon: Post 24 August 2006 JROC*, August 2006, pp. 42–43.

ETM subtasks. International security forces, intelligence support, coordination with indigenous security forces, international civilian-military coordination.

Participate in Stability Operations-Related Regional Security Arrangements

Working definition. The capability to negotiate, participate in, and comply with regional security arrangements. This includes arrangements that enhance border security and control as well as those which enhance regional security.

ETM subtasks. Regional security arrangements.

Explanation. Regional security arrangements were separated out because they seemed qualitatively different from the other subtasks that directly focused on coordinating operational elements directly involved in the stability operation.

Establish Representative, Effective Governance and the Rule of Law

This MME's objective is to establish and maintain the institutions and processes required for representative and effective local and national governance that the indigenous population accepts as legitimate.[14] The development of effective governing institutions is a key requirement for establishing government legitimacy and is important for establishing lasting stability.[15] This includes the existence of meaningful avenues of public participation and oversight, substantive separation of powers through institutional checks and balances, and governmental transparency and integrity, a key component of government effectiveness and political stability.[16] These are primarily civilian-led tasks.

---

[14] Derived from the SSTR operations JOC definition for establishing representative, effective government and the rule of law. DoD, *Military Support to SSTR Operations JOC*, Version 2.0, December 2006, p. 61.

[15] HQDA, *Counterinsurgency*, FM 3-24, December 2006, p. 5-15.

[16] Derived from the Governing Justly and Democratically program area 2, Good Governance, definition in *Foreign Assistance Standardized Program Structure and Definitions*, 20 October 2006.

National Constituting Processes

Working definition. None.

ETM subtasks. National dialogue, constitution.

Explanation. This sectoral task involves the processes required for creating a new government structure and for generating national dialogue in support of that process. These tasks are considered to be organic to the capabilities below and therefore will not be treated as a separate required capability.

Establish a Temporary Civil Administration

This capability encompasses the State Department ETM's sectoral requirement to establish transitional governance.

Working definition. The capability to establish an effective temporary civil administration until an effective indigenous or local government can be constituted.[17]

ETM subtasks. International transitional administration, national transitional administration.

Establish Executive Authority

Working definition. The capability to establish, develop, and maintain executive offices, ministries, and independent governmental bodies that operate efficiently and effectively, incorporate democratic principles, are responsive to the public, are accountable, and which can implement and enforce laws, regulations, and policies.[18] This includes the capability to develop and maintain an open, skilled, transparent, professional, and accountable civil service.

ETM subtasks. Executive mandate and structure, civil service staffing, revenue generation and management, government resources and facilities.

---

[17] Derived from Army Tactical Task 6.16.6, Establish Temporary Civil Administration. HQDA, *Army Universal Task List,* FM 7-15, July 2006, p. 6-120; HQDA, *Civil Affairs Operations,* FM 41-10, February 2000, pp. 2-27 to 2-33, G-3.

[18] Derived from the Governing Justly and Democratically program element 2.2, Public Sector Executive Function, definition in *Foreign Assistance Standardized Program Structure and Definitions,* 20 October 2006.

Establish, Develop, and Maintain Legislatures and Legislative Processes

Working definition. The capability to establish, develop, and maintain legislatures and legislative processes that uphold democratic practices, which can produce effective legislation and regulations, are responsive to the populace, encourage public participation in policymaking, hold themselves and the executive branch accountable, and can oversee the implementation of government programs, budgets, and laws.[19]

ETM subtasks. Mandate, citizen access, staffing and training, resources, and facilities.

Assist Local Governance

Working definition. The capability to assist subnational governments to effectively plan, manage, finance, deliver, and account for local public goods and services.

ETM subtasks. Local governance mandate, staffing and training, services, resources, and facilities.

Enhance Transparency and Anti-Corruption

Working definition. The capability to make transparent and accountable government institutions, processes, and policies. It includes the capability to enforce anti-corruption laws and regulations.[20]

ETM subtasks. Anti-corruption, oversight.

Capability to Conduct Legitimate Elections

Working definition. The capability to conduct democratic elections that are a legitimate contestation of ideas and political power and which reflect the will of the people. This includes the capability to establish, develop, and maintain a legal and regulatory framework

---

[19] Derived from the Governing Justly and Democratically program element 2.1, Legislative Function and Processes, definition in *Foreign Assistance Standardized Program Structure and Definitions*, 20 October 2006.

[20] Derived from the Governing Justly and Democratically program area 4, Civil Society, definition in *Foreign Assistance Standardized Program Structure and Definitions*, 20 October 2006.

which allows political parties and entities to operate within a competitive multiparty system.[21]

ETM subtasks. Elections planning and execution, elections monitoring, elections outreach.

Help Establish, Develop, and Sustain Viable Political Parties

Working definition. The ability to help establish, develop, and sustain viable political parties and political entities that are effective and accountable, that represent and respond to citizens' interests, and that govern responsibly and effectively.[22]

ETM subtasks. Party formation, party training.

Build a Civil Society

Working definition. The capability of citizens to freely organize, advocate, and communicate with their government and with each other.[23]

ETM subtasks. Civil society environment, civic education, strengthening civil society capacity and partnerships.

Build a Free Media

Working definition. The capability to establish, develop, and sustain a broadly functioning independent media sector that can reinforce and foster democratic governance.[24]

ETM subtasks. Media professionalism and ethics, media business development, media environment.

---

[21] Derived from the Governing Justly and Democratically program element 3.2, Elections and Political Processes, definition in *Foreign Assistance Standardized Program Structure and Definitions*, 20 October 2006.

[22] Derived from the Governing Justly and Democratically program element 3.3, Democratic Political Parties, definition in *Foreign Assistance Standardized Program Structure and Definitions*, 20 October 2006.

[23] Derived from the Governing Justly and Democratically program element 2.4, Anti-Corruption Reforms, definition in *Foreign Assistance Standardized Program Structure and Definitions*, 20 October 2006.

[24] Derived from the Governing Justly and Democratically program element 4.2, Media Freedom and Freedom of Information, definition in *Foreign Assistance Standardized Program Structure and Definitions*, 20 October 2006.

Justice and Reconciliation

The DOS ETM established three primary temporal goals for the justice and reconciliation technical sector. These goals are to develop mechanisms for addressing past and ongoing grievances, to initiate the construction of a legal system and process for national reconciliation, and to sustain a functioning legal system accepted as legitimate and based on international norms.[25] This technical sector falls into two broad task areas: justice, which relates to the enforcement and administration of civil justice, and reconciliation, which relates to the resolution of past grievances and humanitarian rights violations. When mapped to the DoD MMEs, these tasks fall under the Establish Representative, Effective Governance and the Rule of Law MME.

Justice

Working definition. The primary goal of this task area is to ensure that all persons, institutions and entities, public and private, including the state, are accountable to laws that are publicly promulgated, equally enforced, and independently adjudicated, and which are consistent with international human rights law. This requires measures to ensure adherence to the principles of supremacy of law, equality before the law, accountability to the law, fairness in the application of the law, separation of powers, participation in decisionmaking, legal certainty, avoidance of arbitrariness and procedural and legal transparency.[26] The ultimate goal is the establishment and operation of a self-sustaining public law-and-order system that operates in accordance with internationally recognized standards and with respect to internationally recognized human rights and freedoms.[27]

---

[25] DOS, *Post-Conflict Reconstruction Essential Tasks*, April 2005, p. V-1.

[26] Derived from the Governing Justly and Democratically program subelement 1.0, Rule of Law and Human Rights, definition in *Foreign Assistance Standardized Program Structure and Definitions*, 20 October 2006.

[27] JCS, *Peace Operations,* JP 3-07.3, October 2007.

Provide an Interim Criminal Justice System

Working definition. In the uncertain aftermath of a natural disaster, man-made disaster, or conflict, the ability to provide a functional criminal justice system capable of sustaining law and order until an indigenous capacity to do so has been developed or restored.[28]

ETM subtasks. Interim international criminal justice personnel: judges, prosecutors, defense advocates, court administrators, corrections staffs, police/investigators, and interim international legal code, organized crime, and law enforcement operations.

Provide Judicial Personnel and Infrastructure

Working definition. The capability to establish, develop, and maintain an effective, accountable, and procedurally fair civil and criminal justice institution as well as provide the personnel required for its operation. The system should be capable of ensuring equality before the law by conducting and fair trials. The justice system includes prosecutors, forensic experts, judges, court personnel, public defenders, private bar, law schools, legal professional associations, and training institutions for justice system personnel.[29]

ETM subtasks. Vetting and recruitment, training/mentoring, judicial support facilities, citizen access.

Indigenous Police

Working definition. Moved to Safe and Secure Environment MME.

ETM subtasks. Indigenous police personnel, essential police facilities, accountability/oversight.

Explanation. This capability is inherently tied to security and has been moved to the security technical sector because of the key role that police forces play in ensuring security in a stability operations environment.

---

[28] Derived from JCS, *Peace Operations,* JP 3-07.3, October 2007.

[29] Derived from the Governing Justly and Democratically program element 1.3, Justice System, definition in *Foreign Assistance Standardized Program Structure and Definitions,* 20 October 2006.

Establish, Maintain, and Operate a Fair, Transparent, and Accountable Corrections System

Working definition. The ability to establish, maintain, and operate a fair, transparent, and accountable corrections system that complies with international human rights standards.[30]

ETM subtasks. Incarceration and parole, corrections facilities, training.

Foster Legal System Reform

Working definition. The ability to develop and sustain a democratically derived legal and regulatory framework that is consistent with international human rights standards.[31]

ETM subtasks. Legal system reorganization, code and statutory reform, participation, institutional reform.

Enforce Property Rights

Working definition. The ability to establish or improve transparent, equitable, and accountable institutions that resolve property disputes and enforce property rights.[32]

ETM subtasks. Prevent property conflicts.

Safeguard Human Rights

Working definition. The capability to protect, promote, and enforce internationally recognized human rights standards.[33]

ETM subtasks. Abuse prevention, capacity building, monitoring.

---

[30] Derived from the Governing Justly and Democratically program element 1.3, Justice System, and Peace and Security program subelement 3.1.2, Corrections Assistance, definitions in *Foreign Assistance Standardized Program Structure and Definitions*, 20 October 2006.

[31] Derived from the Governing Justly and Democratically program element 1.1 Constitutions, Laws and Legal Systems definition in *Foreign Assistance Standardized Program Structure and Definitions*, 20 October 2006.

[32] Derived from the Economic Growth program subelement 6.1.1, Property Rights, definition in *Foreign Assistance Standardized Program Structure and Definitions*, 20 October 2006.

[33] Derived from Governing Justly and Democratically program subelement 1.4, Human Rights, in *Foreign Assistance Standardized Program Structure and Definitions*, 20 October 2006.

Conduct Programs to Combat Human Trafficking

Working definition. The ability to develop, execute, and sustain anti-trafficking programs and to provide support for and protection of trafficking victims.[34] This ability includes supporting and protecting the victims of trafficking, developing legislation that allows for the prosecution of human traffickers, and public awareness campaigns that seek to prevent trafficking.

ETM subtasks. Anti-trafficking strategy, assistance for victims, anti-trafficking legislation.

Support Reconciliation

Working definition. The primary purpose of reconciliation is to address past human rights abuses and social traumas through legal procedures that build respect for the rule of law. It is also intended to promote justice, psychological relief, and reconciliation in order to achieve a sustainable peace.[35]

Address Past War Crimes and Human Rights Violations

Working definition. The ability to address past war crimes and human rights violations through retributive justice mechanisms such as war crimes courts and tribunals which are transparent, accountable, and conform to international legal norms.[36] This includes the establishment and operation of courts and tribunals, the investigation of alleged crimes, the arrest and detention of suspected criminals, and the public dissemination of court records and results.

ETM subtasks. Establishment of courts and tribunals, investigation and arrest, citizen outreach.

---

[34] Derived from the Peace and Security program element 5.3, Trafficking-In-Persons and Migrant Smuggling, definition in *Foreign Assistance Standardized Program Structure and Definitions*, 20 October 2006.

[35] Derived from JCS, *Peace Operations,* JP 3-07, October 2007, p. IV-8.

[36] Derived from the Governing Justly and Democratically program subelement 1.1.3, Transitional Justice, definition in *Foreign Assistance Standardized Program Structure and Definitions*, 20 October 2006.

Establish Truth Commissions and Support Remembrance

Working definition. The ability to address past war crimes and human rights violations through restorative justice mechanisms such as truth and reconciliation commissions and reparations.[37] This includes the establishment and operation of truth commissions and the ability to establish and execute reparations programs.

ETM subtasks. Truth commission organization, reparations, public outreach.

Support Community Rebuilding

Working definition. The ability to provide the local populace with the means to form a cohesive society.[38]

ETM subtasks. Ethnic and intercommunity confidence building, religion and customary justice practices, assistance to victims and remembrance, women, vulnerable populations, evaluating and learning.

Deliver Humanitarian Assistance

The objective of this MME is to rapidly relieve or reduce the results of natural or man-made disasters or other endemic conditions such as human suffering, disease, or privation that might represent a serious threat to life or that can result in great damage to or loss of property through the delivery of humanitarian assistance.[39] Such operations are intended to be emergency in nature, and while they should help create the foundations for long-term recovery and development, they are not a substitute for the development investments required to reduce chronic poverty or establish social services.[40] The effective delivery of humani-

---

[37] Derived from the Governing Justly and Democratically program subelement 1.1.3, Transitional Justice, definition in *Foreign Assistance Standardized Program Structure and Definitions*, 20 October 2006.

[38] HQDA, *Stability Operations*, FM 3-07, October 2008, p. 3-9.

[39] JCS, *Peace Operations,* JP 3-07.3, October 2007, p. IV-5; DoD, *Military Support to SSTR Operations JOC*, Version 2.0, December 2006, p. 42.

[40] Derived from the Humanitarian Assistance definition in *Foreign Assistance Standardized Program Structure and Definitions*, 20 October 2006.

tarian assistance requires the ability to obtain and redistribute essential supplies, food, and medicine within an affected region, or deliver essential items that are not available locally or regionally, to the disaster sites.[41]

We also included humanitarian de-mining in this MME because it is a well-recognized post-conflict humanitarian assistance task and is important for the safety and well-being of the indigenous population. In addition, the State Department ETM includes this mission within the Humanitarian Assistance and Social Well Being technical sector.

The required capabilities for this MME are drawn primarily from the Stabilization, Security, and Reconstruction Operations Joint Operating Concept, but also include elements from the DOS ETM.

Conduct Refugee and Internally Displaced Person Operations

Working definition. The ability to plan, construct, and operate camps and facilities for refugees and internally displaced persons.[42]

ETM subtasks. Prevention of population displacements, refugee assistance, assistance for internally displaced persons, and security for refugees and internally displaced person camps.

Provide Emergency Power Supply

Working definition. The ability to promptly deliver, operate, and maintain electrical power generation equipment to affected regions.[43]

Provide Emergency Water Supply and Sanitation Services

Working definition. The ability to promptly deliver, operate, and maintain emergency water purification, water distribution systems, and to meet basic sanitation standards in the affected regions.[44]

---

[41] DoD, *Military Support to SSTR Operations JOC*, Version 2.0, December 2006, p. 59.

[42] Derived from DoD, *Military Support to SSTR Operations JOC*, Version 2.0, December 2006, p. 60; and DOS, *Post-Conflict Reconstruction Essential Tasks*.

[43] Derived from DoD, *Military Support to SSTR Operations JOC*, Version 2.0, December 2006, p. 59.

[44] Derived from DoD, *Military Support to SSTR Operations JOC*, Version 2.0, December 2006, p. 59, and DOS, Humanitarian Assistance program subelement 1.2.2, Water and Sanitation Commodities and Services, definition in *Foreign Assistance Standardized Program Structure and Definitions*, 20 October 2006.

Provide Emergency Food and Nonfood Supplies Relief

Working definition. The ability to promptly deliver and distribute emergency food and nonfood supplies to affected regions.[45]

ETM subtasks. Famine prevention, emergency food relief, food market response, nonfood relief distribution.

Provide Emergency Shelter in Place

Working definition. The ability to plan and execute emergency shelter programs and to deliver the required supplies in the affected regions [46]

ETM subtasks. Shelter construction.

Explanation. This task is based on the DOS ETM humanitarian affairs shelter construction sectoral subtask and was retained because it is considered to be sufficiently different from the construction and operation of refugee/internally displaced persons camps to warrant separate treatment. It relates to providing emergency shelter to people who do not leave their homes and are also not in organized refugee and internally displaced persons camps.

Provide Emergency Medical Treatment

Working definition. The ability to provide timely emergency medical treatment and prophylaxis to people impacted by natural or man-made disasters.[47]

Conduct Humanitarian De-mining Operations

Working definition. The ability to completely remove all mines and unexploded ordnance after the end of hostilities in order to safeguard the civilian population within a geopolitical boundary.[48] This

---

[45] Derived from the DOS ETM essential task for shelter construction.

[46] Derived from the DOS ETM essential task for shelter construction.

[47] DoD, *Military Support to SSTR Operations JOC*, Version 2.0, December 2006, p. 60.

[48] Derived from the definition for de-mining in FM 20-32. HQDA, *Mine/Countermine Operations*, FM 20-32, October 2002, pp. 9-2, 9-7. See also Peace and Security program subelement 3.4, Explosive Remnants of War, in *Foreign Assistance Standardized Program Structure and Definitions*, 20 October 2006. JP 3-07 categorizes humanitarian de-mining as a security function. JCS, *Peace Operations*, JP 3-07.3, October 2007, p. IV-4.

includes programs to educate the civilian population as to the dangers of mines and unexploded ordnance as well programs to assist and rehabilitate civilians injured by mines and other ordnance.

ETM subtasks. Mine awareness, mine detection, mine clearance, survivor assistance.

Reconstruct Critical Infrastructure and Restore Essential Services

The object of this MME is to address the life support needs of the indigenous population. In an unstable environment, the U.S. military may initially have the lead role in this task, as other agencies may not be present or may lack the capability and capacity to meet the needs of the indigenous population. Due to uncertainties in the security environment, the military must be prepared to perform these tasks for an extended period and under difficult security circumstances.[49] These operations are conducted to prevent the loss of life and the spread of insurgency.[50] The essence of these capabilities is the ability to operate in an uncertain and potentially hostile environment in the aftermath of a disaster (natural or man-made) or conflict. This is emergency or crisis support, not traditional developmental aid, although it may serve as the foundation for such developments.

Traditionally, the reconstitution of the police force has been considered an integral part of the process of essential service restoration. However, given the importance of police forces for the establishment and maintenance of a secure environment, we considered police to be a security-related task rather than a humanitarian task.

Restore, Establish, and Maintain Firefighting Services

Working definition. In the uncertain aftermath of a natural disaster, man-made disaster, or conflict, the ability to restore, establish, and

---

[49] HQDA, *Counterinsurgency,* FM 3-24, 2006, pp. 5-14 to 5-15.

[50] JCS, *Peace Operations,* JP 3-07.3, October 2007, p. IV-9.

maintain firefighting services capable of a timely response to property fires.[51]

ETM subtasks. None.

Explanation. The DOS ETM list does not specifically list the establishment or restoration of firefighting services as an essential task. It is, however, considered an essential service by FM 3-24.[52]

Build, Restore, Maintain, and Operate Water Purification Plants and Potable Water Distribution Systems

Working definition. In the uncertain aftermath of a natural disaster, man-made disaster, or conflict, the ability to build, restore, maintain, and operate water purification plants and potable water distribution systems.[53] The primary objective of this ability is to ensure that water treatment plants and the distribution systems for potable water are functional.[54]

ETM subtasks. Potable water management.

Build, Restore, Maintain, and Operate Power Generation Grids

Working definition. In the uncertain aftermath of a natural disaster, man-made disaster, or conflict, the ability to build, restore, maintain, and operate power generation grids and to ensure the local distribution of electrical power.[55] The primary objective of this task is to ensure that power plants are operational and power lines are intact and functioning.[56]

ETM subtasks. None.

Explanation. The DOS ETM list does not specifically list the restoration of electrical power as an essential task. However, FM 3-24

---

[51] Derived from HQDA, *Counterinsurgency*, FM 3-24, 2006, p. 5-15.

[52] HQDA, *Counterinsurgency*, FM 3-24, 2006, p. 5-15.

[53] Derived from DoD, *Military Support to SSTR Operations JOC*, Version 2.0, December 2006, p. 60.

[54] HQDA, *Counterinsurgency*, FM 3-24, 2006, p. 5-15.

[55] Derived from DoD, *Military Support to SSTR Operations JOC*, Version 2.0, December 2006, p. 60; and DOS, *Post-Conflict Reconstruction Essential Tasks*, April 2005, pp. IV-15 to IV-16.

[56] HQDA, *Counterinsurgency*, FM 3-24, 2006, p. 5-15.

and the *Military Support to Stabilization, Security, and Reconstruction Operations Joint Operating Concept* consider the provision of electricity to be an essential service.

Build, Restore, Maintain, and Operate Schools and Universities

Working definition. In the uncertain aftermath of a natural disaster, man-made disaster, or conflict, the ability to build, restore, maintain, and operate schools and universities.[57] The primary objective of this capability is to ensure that schools and universities are open, staffed, and supplied.[58]

ETM subtasks. Human resources, education—schools, education—universities, curriculum, literacy campaign.

Repair and Maintain Transportation Networks

Working definition. In the uncertain aftermath of a natural disaster, man-made disaster, or conflict, the ability to repair, construct, maintain, and operate roads, bridges, tunnels, ports, and airfields for road, rail, air, and sea transportation.[59] The primary objective of this capability is to ensure that transportation networks are open and trafficable.[60]

ETM subtasks. Assess condition of existing transportation facilities, construct expedient repairs, or build new transportation facilities to support security and stabilization and to facilitate re-establishment of commerce.[61]

Explanation. See also the Economic Development MME.

Repair and Maintain Public Health Facilities

Working definition. In the uncertain aftermath of a natural disaster, man-made disaster, or conflict, the ability to repair, build, maintain,

---

[57] Derived from DoD, *Military Support to SSTR Operations JOC*, Version 2.0, December 2006, p. 60; and DOS, *Post-Conflict Reconstruction Essential Tasks*, April 2005, p. III-10.

[58] HQDA, *Counterinsurgency*, FM 3-24, 2006, p. 5-15.

[59] Derived from DoD, *Military Support to SSTR Operations JOC*, Version 2.0, December 2006, p. 60.

[60] Derived from HQDA, *Counterinsurgency*, FM 3-24, 2006, p. 5-15.

[61] Derived from DOS, *Post-Conflict Reconstruction Essential Tasks*, April 2005, pp. IV-15 to IV-16.

and operate primary health care clinics, hospitals, and other elements of the health care system.[62] The primary objective of this capability is to ensure that hospitals and clinics are open and staffed.[63] This ability includes the provision of ambulance services.

ETM subtasks. Medical capacity, local public health clinics, hospital facilities, human resources development for health care workforce, health policy and financing, prevention of epidemics, HIV/AIDS, nutrition, reproductive health, environmental health, community health education.

Maintain Public Sanitation

Working definition. In the uncertain aftermath of a natural disaster, man-made disaster, or conflict, the ability to repair, construct, maintain, and operate sewage disposal systems and collect and dispose of garbage.[64] The primary objective of this ability is to ensure that trash is collected regularly and that the sewage system is operating.[65]

ETM subtasks. Sanitation and wastewater management.

Build, Restore, Maintain, and Operate Telecommunication Networks

Working definition. In the uncertain aftermath of a natural disaster, man-made disaster, or conflict, the ability to build, restore, maintain, and operate telecommunication networks.[66]

ETM subtasks. None.

Explanation. Neither the DOS ETM list nor FM 3-24 specifically identifies the restoration of telecommunications networks as an

---

[62] Derived from DoD, *Military Support to SSTR Operations JOC*, Version 2.0, December 2006, p. 60; and S/CRS, *Post-Conflict Reconstruction Essential Tasks*, April 2005, pp. III-8 to III-10.

[63] HQDA, *Counterinsurgency*, FM 3-24, 2006, p. 5-15.

[64] Derived from DoD, *Military Support to SSTR Operations JOC*, Version 2.0, December 2006, p. 60.

[65] HQDA, *Counterinsurgency*, FM 3-24, 2006, p. 5-15.

[66] Derived from DoD, *Military Support to SSTR Operations JOC*, Version 2.0, December 2006, p. 60.

essential task or service.[67] It is, however, identified as a critical enabling capability by the *Military Support to Stabilization, Security, and Reconstruction Operations Joint Operating Concept.*[68] See also the Economic Development MME.

Social Well-Being

This DOS ETM Technical Sector has been dropped as a separate category and its functions moved elsewhere or absorbed into existing essential capabilities.

Trafficking in Persons

Explanation. This DOS sectoral task does not fit well in any of the "humanitarian" categories, so we have moved it to the Reconstruct Critical Infrastructure and Restore Essential Services MME because its main focus is primarily a law enforcement issue.

Public Health

Explanation. Moved to Reconstruct Critical Infrastructure and Restore Essential Services MME.

Education

Explanation. Moved to Reconstruct Critical Infrastructure and Restore Essential Services MME.

Social Protection

Explanation. This task was subsumed into a variety of other capabilities, the three most important being the social safety net sectoral task in the Reconstruct Critical Infrastructure and Restore Essential Services MME and the human rights and community rebuilding sectoral tasks in the Establish Representative, Effective Governance and the Rule of Law MME.

---

[67] Telecommunications systems are included in the ETM list as part of the Infrastructure task area, but the focus appears to be more on assessment and long-term development rather that the meeting of immediate communications requirements.

[68] DoD, *Military Support to SSTR Operations JOC*, December 2006, p. 60.

Assessment, Analysis, and Reporting

Explanation. This sectoral task has been eliminated because it is considered to be organic to the capabilities to conduct the above missions. Assessment, analysis, and reporting are deemed to be an integral part of a capability.

Support Economic Development

The primary goal of this MME is to promote economic development that addresses near-term problems such as large-scale unemployment and the re-establishment of economic activity in a way that lays the foundation for sustained economic growth that stimulates indigenous economic activity. A viable economy is a key component of stability and reinforces government legitimacy.[69]

The DOS ETM established three primary temporal goals for the economic stabilization and infrastructure technical sector. These goals are to respond to immediate needs, to set the conditions for further development, and to institutionalize a long-term development program.[70]

The Joint Publication (JP) 3-07.3 notes that the military may need to start restoring economic infrastructure in the absence of civilian agencies. JP 3-07.3 lists the reconstitution of power, transportation, communications, health and sanitation, firefighting, mortuary services, and environmental control.[71] Therefore, we have separated out economic infrastructure tasks from the Reconstruct Critical Infrastructure and Restore Essential Services MME and placed them here under the Support Economic Development MME.

---

[69] Derived from the Army and DoD definitions for supporting economic development. HQDA, *Counterinsurgency,* FM 3-25, December 2006; DoD, *Military Support to SSTR Operations JOC,* December 2006, pp. 43–44.

[70] DOS, *Post-Conflict Reconstruction Essential Tasks,* April 2005, p. IV-1.

[71] JCS, *Peace Operations,* JP 3-07.3, October 2007.

Economic Stabilization

JP 3-07.3 defines economic stabilization as primarily a civilian responsibility that focuses on restoring employment opportunities, initiating market reforms, mobilizing foreign and domestic investment, supervising monetary reform, and rebuilding public structures. It does note, however, that the military must be prepared to undertake these tasks when civilian agencies are absent.[72]

Generate Employment

Working definition. The ability to design, fund, and implement public works initiatives, to stimulate micro and small enterprise, and foster workforce development programs that will rapidly provide employment for the indigenous population.[73]

ETM subtasks. Public works jobs, micro and small enterprise stimulation, skills training, and counseling.

Develop Monetary Policy

Working definition. The ability to develop mechanisms and institutions, including the ability to set and control interest rates that allow the government to manage the economy by expanding or contracting the money supply.[74]

ETM subtasks. Central bank operations, macroeconomic policy and exchange rates, monetary audit, monetary statistics.

Develop and Apply Fiscal Policy and Governance

Working definition. The ability to develop and apply sustainable, efficient, and transparent fiscal policies that can generate the resources required to sustain key public functions. This includes the ability to establish revenue and expenditure structures and to manage the economy through the expansion and contraction of government spending and to design and administer public expenditure systems that are

---

[72] JCS, *Peace Operations,* JP 3-07.3, October 2007.

[73] Derived from DoD, *Military Support to SSTR Operations JOC*, Version 2.0, 2006, p. 61; and the DOS essential tasks for this sectoral task.

[74] Derived from the Economic Growth program element 1.2, Monetary Policy, definition in *Foreign Assistance Standardized Program Structure and Definitions*, 20 October 2006.

transparent and which lend themselves to the equitable and timely formulation of budgets and which can plan for the needs of the entire population.[75]

ETM subtasks. Fiscal and macroeconomic policy, treasury operations, budget, public sector investment, revenue generation, tax administration, customs reform, enforcement, tax policy, fiscal audit.

Promote General Economic Policies

Working definition. None.

ETM subtasks. Strategy/assessment, prices and subsidies, international financial assistance—donor coordination, public sector institutions.

Establish, Develop, Regulate, and Sustain a Well-Functioning and Equitable Financial Sector

Working definition. The ability to establish, develop, regulate, and sustain a well-functioning and equitable financial sector.[76]

ETM subtasks. Banking operations, banking regulations and oversight, banking law, bank lending, asset and money laundering, non-banking sector, stock and commodity markets.

Manage and Control Both Foreign and Domestic Borrowing and Debt

Working definition. The ability to manage and control both foreign and domestic borrowing and debt.[77]

ETM subtasks. Debt management, arrears clearance.

Develop Trade

Working definition. The ability to establish, develop, sustain, and enforce trade policies, laws, regulations, and administrative practices

---

[75] Derived from the Economic Growth program element 1.1, Fiscal Policy, definition in *Foreign Assistance Standardized Program Structure and Definitions*, 20 October 2006.

[76] Derived from the Economic Growth program area 3, Financial Sector, definition in *Foreign Assistance Standardized Program Structure and Definitions*, 20 October 2006.

[77] Derived from the Economic Growth program subelement 1.2.5, Debt Management, definition in *Foreign Assistance Standardized Program Structure and Definitions*, 20 October 2006.

that support improvement in the trade environment and which facilitate international trade.[78]

ETM subtasks. Trade structure, trade facilitation.

Promote a Market Economy

Working definition. The ability to support the establishment or re-establishment of a functioning market economy.

ETM subtasks. Private sector development, small and microenterprise regime, privatization, natural resources and environment.

Promote Legal and Regulatory Reform

Working definition. The ability to support the development of a legal and regulatory framework supportive of a market economy.

ETM subtasks. Property rights, business/commercial law, labor, economic legal reform, competition policy, public utilities and resources regulation, economic enforcement and anti-corruption.

Promote Agricultural Development

Working definition. The ability to support the establishment or re-establishment of a viable agricultural sector capable of long-term growth.[79]

ETM subtasks. Agricultural land and livestock, agricultural inputs, agricultural policy and financing, agricultural distribution.

Establish a Social Safety Net

Working definition. The ability to support the establishment of social safety net programs.

ETM subtasks. Pension system, social entitlement funds, women's issues.

Economic Infrastructure

The main objective of this sub-MME is to support the creation, improvement, and sustainability of physical infrastructure and related

---

[78] Derived from the Economic Growth program element 2.1, Trade and Investment Enabling Environment, definition in *Foreign Assistance Standardized Program Structure and Definitions*, 20 October 2006.

[79] HQDA, *Stability Operations*, FM 3-07, October 2008, p. 3-17.

services in order to enhance the economic environment and to improve economic productivity. The main economic infrastructure elements are transportation, telecommunications, and energy.[80]

The joint publication JP 3-07.3 notes that the military may need to start restoring economic infrastructure in the absence of civilian agencies. JP 3-07.3 lists the reconstitution of power, transportation, communications, health and sanitation, firefighting, mortuary services, and environmental control.[81] Therefore, we have separated out economic infrastructure tasks from the Reconstruct Critical Infrastructure and Restore Essential Services MME.

Build and Maintain Transportation Infrastructure

Working definition. The ability to design, execute, and sustain investment and regulatory programs that support and strengthen reliable and affordable transport systems, including roads, airports, railways, and ports.[82]

ETM subtasks. Transportation sector policy and administration, airports infrastructure, roads infrastructure, railway infrastructure, ports and waterway infrastructure.

Explanation. See also Reconstruct Critical Infrastructure and Restore Essential Services MME.

Develop, Strengthen, and Support Telecommunications Infrastructure

Working definition. The ability to develop, strengthen, and support communications networks through investment and regulatory reform.[83]

---

[80] Derived from the Economic Growth program area 4, Infrastructure, definition in *Foreign Assistance Standardized Program Structure and Definitions*, 20 October 2006.

[81] DoD, *Peace Operations,* JP 3-07.3 Request for Comment Draft, June 2006, p. IV-10.

[82] Derived from the Economic Growth program element 4.3, Transport Services, definition in *Foreign Assistance Standardized Program Structure and Definitions*, 20 October 2006.

[83] Derived from the Economic Growth program element 4.2, Communications Services, definition in *Foreign Assistance Standardized Program Structure and Definitions*, 20 October 2006.

ETM subtasks. Telecommunications policy and administration, telecommunication infrastructure.

Explanation. See also Essential Services MME.

Develop and Maintain Energy Infrastructure

Working definition. The ability to develop, execute, and sustain programs that increase the efficiency and reliability of energy services and which promote investment in the development, transport, processing, and utilization of indigenous energy sources and imported fuels.[84] Of particular importance is the ability to develop the production and distribution of fossil fuels and the generation and distribution of electrical power.

ETM subtasks. Fossil fuels production and distribution, electrical power sector, energy infrastructure.

Explanation. See also Reconstruct Critical Infrastructure and Restore Essential Services MME.

Build and Maintain General Infrastructure

Working definition. The ability to develop, execute, and sustain general infrastructure programs that promote overall and municipal indigenous governance, commerce, and social well-being.[85]

ETM subtasks. Engineering and construction, municipal services.

Explanation. See also Reconstruct Critical Infrastructure and Restore Essential Services MME.

Conduct Strategic Communications

The primary goal of this MME is to communicate effectively to key local and foreign audiences information regarding the stability opera-

---

[84] Derived from the Economic Growth program element 4.1, Modern Energy Services, definition in *Foreign Assistance Standardized Program Structure and Definitions*, 20 October 2006.

[85] Derived from the DOS essential tasks for this sectoral task.

tion in order to preserve conditions favorable to achieving the overall SSTR operation goals and objectives.[86]

This task is included separately within each of the State Department ETM technical sectors. Military doctrine and concepts, however, tend to treat strategic communications/information operations as a separate MME or LLO that cross-cuts all the other MMEs and LLOs.[87] In light of this, we have pulled the public information and communications sectoral tasks out of the individual DOS ETM technical sectors and combined them into a broad single required capability.

Conduct Public Information and Communication Activities

Working definition. The ability to plan, coordinate, and synchronize public information activities and resources to support the objectives of the stability operation through the communication of truthful, timely, and factual unclassified information within the area of operations to foreign, domestic, and internal audiences.[88] This capability includes the expansion, development, or establishment of indigenous media outlets and the training of indigenous journalists.

ETM subtasks. Disseminate information on all MMEs.

---

[86] DoD, *Military Support to SSTR Operations JOC*, Version 2.0, December 2006, pp. 61–62.

[87] See DoD, *Military Support to SSTR Operations JOC*, Version 2.0, December 2006, pp. 20–21; and HQDA, *Full Spectrum Operations (DRAG Draft)*, FM 3-0, November 2006, p. 6-13; HQDA, *Counterinsurgency*, FM 3-24, December 2006, pp. 5-3 to 5-6.

[88] Derived from the Joint Public Affairs Operations tier 1 joint capability area definition. *Joint Capability Areas Tier 1 and Supporting Tier 2 Lexicon: Post 24 August 2006 JROC*, August 2006, p. 37. An alternative definition derived from the stability operations JOC is "the capability to conduct effective strategic communications that engage key local and foreign audiences in order to create, strengthen, or preserve conditions favorable to the achievement of overall SSTR goals and objectives." DoD, *Military Support to SSTR Operations JOC*, Version 2.0, December 2006, p. 61.

List of BPC for Stability Operations Programs and Activities

Used to support the baseline descriptive analysis in Chapter Three, this appendix contains detailed lists of BPC for stability operations programs and activities managed by the U.S. Army (Table B.1), other DoD organizations (Table B.2), non-DoD U.S. government agencies (Tables B.3 and B.4), and major U.S. allies (Table B.5). In most cases, the lists include the name of the executing agency or country, the program title, the program objective, the DoD stability operations major mission element associated with the program, the geographic focus of the program, and whether the program supports the development of indigenous and/or coalition forces in partner countries.

Table B.1
Army BPC for Stability Operations Programs

Agency	Program Title	Program Objective	Primary MME	Geographic Focus	Indigenous Forces	Coalition Forces
Army (U.S. Army Corps of Engineers)	Civil-Military Emergency Preparedness (CMEP)	Support military commanders in achieving their security objectives by facilitating disaster preparedness workshops to train foreign nationals in international disaster response.	Deliver humanitarian assistance	EUCOM AOR	Yes	Yes
Army	Multilateral Interoperability Program	Achieve international interoperability of command and control information systems at all levels.	Establish and maintain a safe and secure environment	EUCOM AOR, NATO (primarily)	No	Yes
Army	Reciprocal Unit Exchange Program		Establish and maintain a safe and secure environment	Global		
Army	Administrative and Professional Exchange Program	Exchange of military or civilian specialist personnel in administrative, finance, health, legal, logistics, planning, and other support in host organizations.	Establish and maintain a safe and secure environment			
Army	Engineer and Scientist Exchange Program	Improve understanding of the other nation's technical capabilities and the process by which its defense RDT&E program is managed by exchanging military or civilian engineers and scientists in research, development, test and evaluation positions in host organizations.	Establish and maintain a safe and secure environment			

Table B.1 (continued)

Agency	Program Title	Program Objective	Primary MME	Geographic Focus	Indigenous Forces	Coalition Forces
Army	Military Personnel Exchange Program	Foster mutual understanding between the military establishment of each participating nation by providing exchange personnel familiarity with the organization, administration, and operations of the host organization.	Establish and maintain a safe and secure environment	Global	No	Yes
Army	Foreign Liaison Officer Program	Facilitate cooperation, mutual understanding and information exchange regarding concepts or capabilities development, training, doctrine, research and development, operations, etc., between the defense establishments of the United States and our allies and coalition partners.	Establish and maintain a safe and secure environment			
Army	Security assistance-funded medical programs		Deliver humanitarian assistance			
Army	Foreign Visits Program (medical)		Deliver humanitarian assistance			
Army	Personnel Exchange Program (medical)		Deliver humanitarian assistance			
Army	Foreign Liaison Program (medical)		Deliver humanitarian assistance			

Table B.1 (continued)

Agency	Program Title	Program Objective	Primary MME	Geographic Focus	Indigenous Forces	Coalition Forces
Army	Emergency Management International		Deliver humanitarian assistance	CMEP in AORs beyond EUCOM AOR	Yes	Yes
Army	West Point Center for the Rule of Law	Inspire a steadfast commitment to the rule of law; bring together scholars and practitioners to facilitate development of the rule of law.	Establish representative, effective governance and the rule of law	Global	Yes	Yes
Army	The Judge Advocate General's Legal Center and School	Prepare Army personnel and interagency partners for work related to rule of law development.	Establish representative, effective governance and the rule of law	Global	Yes	Yes
Army	Center for Law and Military Operations	Publishes a handbook for deploying rule of law practitioners; serves as a central hub for information related to stability operations.	Establish representative, effective governance and the rule of law	Global	Yes	Yes

Table B.2
Other DoD BPC for Stability Operations Programs

Agency	Program Title	Program Objective	Primary MME	Geographic Focus	Indigenous Forces	Coalition Forces
Office of the Secretary of Defense (OSD)	Defense Resource Management Institute	Help partner countries to develop an understanding and appreciation of the concepts, techniques, and decisionmaking skills related to defense resource management.	Establish and maintain a safe and secure environment			
OSD	Defense Planning Exchange	Host working-level Central and East European military and civilian officials for detailed exchanges in order to familiarize them with how the United States builds a strategy-based, balanced defense program to assist in defense planning and modernization decisions.	Establish and maintain a safe and secure environment			
OSD	Cooperative Threat Reduction Defense and Military Contacts Program	Expand defense and military contacts between the United States and the former Soviet Union to promote U.S. national security objectives in the former Soviet states.	Establish and maintain a safe and secure environment	EUCOM, CENTCOM	Yes	Yes
OSD	Regional Counterterrorism Fellowship Program	Send foreign military officers to U.S. military educational institutions and selected regional centers for non-lethal training.	Establish and maintain a safe and secure environment	Global	Yes	Yes

Table B.2 (continued)

Agency	Program Title	Program Objective	Primary MME	Geographic Focus	Indigenous Forces	Coalition Forces
OSD	Cooperative Threat Reduction Weapons of Mass Destruction Proliferation Prevention Initiative	Bolster non-Russian former Soviet states' ability to prevent proliferation of weapons of mass destruction across their borders.	Establish and maintain a safe and secure environment	EUCOM, CENTCOM	Yes	No
JFCOM J-9	Logistic Information Exchange Program	Improve coalition capabilities on strategic and operational level.	Establish and maintain a safe and secure environment			
COCOMs	Medical Assistance Programs		Deliver humanitarian assistance	Global	Yes	Yes
	Traditional Commander Activities-funded medical programs		Deliver humanitarian assistance			
OSD	Regional centers: George C. Marshall Center	Assist partners to develop surrogate military doctrine for peacekeeping and stability operations.	Establish and maintain a safe and secure environment	Global	Yes	Yes
OSD/ National Defense University	Regional centers: Center for Hemispheric Defense Studies	Advanced stability operations course in development to develop management and leadership skills.	Establish and maintain a safe and secure environment	Global	Yes	Yes
OSD	Regional centers: Asia-Pacific Center for Security Studies	Encourage partners to look at governance and its relationship to security.	Establish representative and effective governance	Global	Yes	Yes

Table B.2 (continued)

Agency	Program Title	Program Objective	Primary MME	Geographic Focus	Indigenous Forces	Coalition Forces
OSD	Regional centers: Near East South Asia Center for Strategic Studies	Offer various courses related to stability operations and run regional workshops for outreach purposes.	Establish and maintain a safe and secure environment	Global	Yes	Yes
OSD	Regional centers: Africa Center for Strategic Studies	Facilitate the ongoing Africa Standby Force workshops to determine outstanding requirements for establishing a regional peacekeeping brigade, working with ECOWAS to build its strategic capacity, working with some of the African War Colleges to build ties to U.S. War Colleges. Offer stability operations course.	Establish and maintain a safe and secure environment	Global	Yes	Yes
DOS	NATO School	Conduct individual, operational level education and training on NATO's current and emerging strategy, concepts, doctrine, policies, and procedures to improve the operational effectiveness of the Alliance.	Establish and maintain a safe and secure environment	EUCOM, CENTCOM	Yes	Yes
OSD	Warsaw Initiative Fund			EUCOM, CENTCOM	Yes	Yes
COCOMs	Joint Contact Team Program			EUCOM	Yes	Yes

Table B.2 (continued)

Agency	Program Title	Program Objective	Primary MME	Geographic Focus	Indigenous Forces	Coalition Forces
Joint Staff	Joint Chiefs of Staff and other Multilateral Exercises		Establish and maintain a safe and secure environment	Global	Yes	Yes
National Guard Bureau	State Partnership Program		Establish and maintain a safe and secure environment	Global	Yes	Yes
Defense Threat Reduction Agency	International Counterproliferation Program		Establish and maintain a safe and secure environment	EUCOM, CENTCOM	Yes	No

Table B.3
Interagency BPC for Stability Operations Programs

Agency	Program Title	Program Objective	Primary MME	Geographic Focus	Indigenous Forces	Coalition Forces
DOS	Office of the Coordinator for Reconstruction and Stabilization	Coordinate U.S. government civilian capacity to promote stability, reconstruction, democracy, and economic development.	Establish and maintain a safe and secure environment	Global	Yes	No
DOS	Bureau of International Narcotics and Law Enforcement Affairs	Minimize entry of illegal drugs into U.S. and impact of international crime.	Establish and maintain a safe and secure environment	Global	Yes	No
DOS	Bureau of Democracy, Human Rights, and Labor	Promote democracy, support newly-formed democracies, and identify and denounce regimes that act counter to democratic principles.	Establish representative and effective governance	Global	Yes	No
DOS	Bureau of Economic, Energy and Business Affairs	Establish fair rules of international trade with the World Trade Organization, negotiate bilateral and regional investment treaties, combat bribery in international commerce, negotiate debt relief, coordinate issues related to economic sanctions, and foster energy security.	Economic stabilization and infrastructure	Global	Yes	No
DOS	Bureau of Political-Military Affairs	Provide policy direction in the areas of international security, security assistance, military operations, defense strategy and policy, military use of space, and defense trade.	Establish and maintain a safe and secure environment	Global	Yes	No

Table B.3 (continued)

Agency	Program Title	Program Objective	Primary MME	Geographic Focus	Indigenous Forces	Coalition Forces
DOS	Bureau of Population, Refugees, and Migration	Coordinate U.S. international population policy and promote its goals through international cooperation to assist and protect refugees abroad.	Deliver humanitarian assistance	Global	Yes	No
DOS	Export Control and Related Border Security Assistance	Provide training to promote foreign capacity to control borders with emphasis on nonproliferation.	Establish and maintain a safe and secure environment	Global	Yes	No
DOS	Enhanced International Peacekeeping	Produce highly skilled peacekeeping trainers who have been introduced to a variety of training and educational methods and have a good grasp of peace support operations policy, doctrine, logistics and interoperability issues.	Establish and maintain a safe and secure environment	Global	Yes	Yes
DOS	International Law Enforcement Academy	Combat international drug trafficking, criminality, and terrorism through strengthened international cooperation.	Establish and maintain a safe and secure environment	Global	Yes	No
USAID	Office of Transition Initiatives	Provide short-term assistance to key countries to support reconciliation, local economies, nascent independent media, and peace and transition to democracy.	Establish representative and effective governance	Global	Yes	No
USAID	Office of U.S. Foreign Disaster Assistance	Provide assistance in targeted sectors (e.g., shelter, health, water, sanitation, nutrition, coordination) to reduce human and economic consequences of disasters. Improve risk management activities to enhance local response capacity.	Deliver humanitarian assistance	Global	Yes	No

Table B.3 (continued)

Agency	Program Title	Program Objective	Primary MME	Geographic Focus	Indigenous Forces	Coalition Forces
USAID	Bureau for Economic Growth, Agriculture, and Trade	Reduce poverty and promote prosperity by supporting economic growth, trade and investment, microenterprise development, urban development, development credit, education, agriculture, natural resource management, science policy, energy, information and communications technology, and technology transfer.	Economic stabilization and infrastructure	Global	Yes	No
USAID	Bureau for Global Health	Promote global health by supporting child, maternal, and reproductive health, and reduce diseases such as HIV/AIDS, malaria, and tuberculosis.	Establish or restore essential services	Global	Yes	No
USAID	Office of Military Affairs	Act as USAID liaison to U.S. and foreign militaries for planning, training, education, and exercises. Develop guidelines and standard operating procedures for interacting with organizations.	Establish and maintain a safe and secure environment	Global	Yes	No
USAID	Office of Conflict Management and Mitigation	Contain or resolve existing or emergent regional conflicts by providing technical support such as conflict assessments and program design.	Establish and maintain a safe and secure environment	Global	Yes	No
USAID	Office of Democracy and Governance	Assist countries in improving governance and transitioning to democracy by supporting judicial, electoral, and civil reform.	Establish representative and effective governance	Global	Yes	No

Table B.3 (continued)

Agency	Program Title	Program Objective	Primary MME	Geographic Focus	Indigenous Forces	Coalition Forces
USAID	Office of Food for Peace	Promote food security with U.S. agricultural resources and processing capabilities to address political instability and environmental degradation.	Deliver humanitarian assistance	Global	Yes	No
USAID	Office of Private Voluntary Coopera-tion/American Schools and Hospitals Abroad	Assist educational and medical institutions in research and training by demonstrating U.S. technologies and practices.	Establish or restore essential services	Global	Yes	No
USAID	Bureau for Asia and the Near East	Promote responsible economic development, health, democracy, and prevention and treatment of HIV/AIDS.	Establish or restore essential services	Regional	Yes	No
USAID	Bureau for Europe and Eurasia	Address international terrorism, cross-border spread of HIV/AIDS, and human trafficking. Promote trade and economic development.	Establish or restore essential services	Regional	Yes	No
USAID	Bureau for Latin America and the Caribbean	Promote responsible economic development, health, democracy, prevention and education of HIV/AIDS, and counternarcotics.	Establish or restore essential services	Regional	Yes	No
USAID	Bureau for Sub-Saharan Africa	Improve access to education and health services, increase the productivity of agriculture, reduce threat from HIV/AIDS pandemic, promote democracy to address violent conflict and instability.	Establish or restore essential services	Regional	Yes	No

Table B.3 (continued)

Agency	Program Title	Program Objective	Primary MME	Geographic Focus	Indigenous Forces	Coalition Forces
DOJ	International Criminal Investigative Training Assistance Program	Support police and corrections forces in emerging democracies, international peacekeeping and post-conflict operations.	Establish and maintain a safe and secure environment	Global	Yes	No
DOJ	Office of Overseas Prosecutorial Development, Assistance and Training	Provide assistance to enhance the capacity of foreign justice sector institutions and personnel to support the rule of law and partner with the U.S. and others in combating terrorism, trafficking in persons and controlled substances, organized crime, corruption, and financial crimes.	Justice and reconciliation	Global	Yes	No
U.S. Inst. of Peace	Center for Post-Conflict Peace and Stability Operations	Promote stability, democracy, economic development, and social reconstruction in societies emerging from conflict.	Establish and maintain a safe and secure environment	Global	Yes	No
U.S. Inst.of Peace	Professional Training Program	Professional training program to improve conflict management skills through all phases of conflict (preventing nascent conflicts, mediating active conflicts, and building peace in conflict's aftermath).	Establish and maintain a safe and secure environment	Global	Yes	No
U.S. Inst.of Peace	Religion and Peacemaking Program	Build the capacity of faith-based and interfaith organizations to be peacemakers in zones of conflict where religion contributes to the conflict.	Establish and maintain a safe and secure environment	Global	Yes	No

Table B.3 (continued)

Agency	Program Title	Program Objective	Primary MME	Geographic Focus	Indigenous Forces	Coalition Forces
U.S. Inst.of Peace	Rule of Law Program	Assist law-based management of international conflict.	Justice and reconciliation	Global	Yes	No
U.S. Inst.of Peace	Virtual Diplomacy Initiative	Explore the role of information and communications technologies in diplomacy, particularly their effect upon international conflict management and resolution.	Establish and maintain a safe and secure environment	Global	Yes	No
USAID	Office of Volunteers for Prosperity	Recruit highly skilled U.S. professionals to support U.S. goals in global health and prosperity.	Establish or restore essential services	Global	Yes	No
DOC	U.S. Patent and Trademark Office	Bring Iraq into compliance with World Trade Organization standards. Hold capacity-building workshops with other countries on intellectual property enforcement issues.	Economic stabilization and infrastructure	Global	Yes	No
DOT	Federal Aviation Administration	Support stability by promoting the development of a sustainable civil aviation system that complies with international standards.	Establish or restore essential services		Yes	No
DOT	Maritime Administration	Support NATO deployment operations and assist in the acquisition of sealift.	Establish and maintain a safe and secure environment	Regional	No	Yes
Treas-ury	Technical Assistance Program	Advance international financial security and stability through focused technical assistance to government agencies in developing and transitioning economies.	Economic stabilization and infrastructure	Global	Yes	No

Table B.3 (continued)

Agency	Program Title	Program Objective	Primary MME	Geographic Focus	Indigenous Forces	Coalition Forces
DHS	Container Security Initiative	Screen containers coming into the United States from foreign countries to detect potential delivery of a terrorist weapon. Prescreen containers overseas before shipped.	Establish and maintain a safe and secure environment	Global	Yes	No
DHS, U.S. Coast Guard	Model Maritime Service Code	Provide technical assistance to partner countries through expert analysis and counsel, equipment transfers, and training.	Establish and maintain a safe and secure environment	Global	Yes	No
DHS, U.S. Coast Guard	U.S. Coast Guard Academy International Cadet Program	Provides training to international participants in leadership, management and technical training provided at Coast Guard schools and operational units in the U.S. and Mobile Training Teams.	Establish and maintain a safe and secure environment	Global	Yes	No
DHS, U.S. Coast Guard	Mobile Training Teams	Provide training and technical assistance to international partners in maritime law enforcement, marine safety/ environmental protection, small boat operations and maintenance, search and rescue, and infrastructure development for countries with waterway law enforcement programs.	Establish and maintain a safe and secure environment	Global	Yes	No
DHS, U.S. Coast Guard	Caribbean Support Tender	Improve the capability of Caribbean nations to operate excess USCG equipment, provide for their own security and safety, and deter the trafficking of drugs, weapons and migrants in the region (terminated).	Establish and maintain a safe and secure environment	Regional	Yes	No

Table B.3 (continued)

Agency	Program Title	Program Objective	Primary MME	Geographic Focus	Indigenous Forces	Coalition Forces
DHS, U.S. Coast Guard	Security Assistance program—Foreign Military Sales program	Sale of excess Coast Guard ship and boat inventory to international partners. Provide equipment, support services, and technical and operational training.	Establish and maintain a safe and secure environment	Global	Yes	No
DHS, U.S. Coast Guard	DEEPWATER	Provide international partners with new-construction vessels and aircraft as well as existing legacy platforms that are made excess by U.S. Coast Guard acquisition of new systems.	Establish and maintain a safe and secure environment	Global	Yes	No
DHS	National Cyber Security Division	Test communications, policies, and procedures in response to various cyber attacks and identify where further planning and process improvements are needed, including implications for physical infrastructure. Exercise interagency and inter-governmental coordination.	Establish and maintain a safe and secure environment	Global	No	Yes
DOE	Office of Nuclear Warhead Protection	Improve the security of the Russian Federation Ministry of Defense nuclear material sites.	Establish and maintain a safe and secure environment	Regional	Yes	No
DOE	Office of Weapon Material Protection	Provide upgrades to nuclear weapons, uranium enrichment, and material processing/storage sites.	Establish and maintain a safe and secure environment	Regional	Yes	No

Table B.3 (continued)

Agency	Program Title	Program Objective	Primary MME	Geographic Focus	Indigenous Forces	Coalition Forces
DOE	Office of Material Consolidation and Civilian Sites	Support upgrade projects at civilian nuclear facilities. Engage countries outside of Russia and the former Soviet Union, including cooperative efforts with China.	Establish and maintain a safe and secure environment	Regional	Yes	No
DOE	Office of National Infrastructure and Sustainability	Establish and sustain effective operation of upgraded systems, and developing strategies for transitioning technical and financial support to the Russian Federation.	Establish and maintain a safe and secure environment	Regional	Yes	No
DOE	Office of Second Line of Defense	Prevent illicit trafficking in nuclear and radiological materials by securing international land borders, seaports, and airports.	Establish and maintain a safe and secure environment	Regional	Yes	No
DOE	International Radiological Threat Reduction Program	Provide training and equipment to locate, identify, recover, consolidate, and enhance the security of radiological materials.	Establish and maintain a safe and secure environment	Regional	Yes	No
EPA	Office of Western Hemisphere and Bilateral Affairs	Implement EPA's bilateral technical assistance, capacity-building, and policy programs with priority countries and regions.	Economic stabilization and infrastructure	Global	Yes	No
EPA	Office of Technology Cooperation and Assistance	Manage international air, water, and toxics programs and for coordinating international training, technology transfer, and environmental information initiatives.	Economic stabilization and infrastructure	Global	Yes	No

Table B.3 (continued)

Agency	Program Title	Program Objective	Primary MME	Geographic Focus	Indigenous Forces	Coalition Forces
EPA	Office of International Environmental Policy	Provides leadership, analysis, and coordination for EPA's work with multilateral organizations areas such as marine pollution and the intersection between international trade and the environment.	Economic stabilization and infrastructure	Global	No	Yes

Table B.4
Iraq/Afghanistan Interagency BPC for Stability Operations Programs

Agency	Program Title	Program Objective	Primary MME	Geographic Focus	Indigenous Forces	Coalition Forces
DOC—ITA	Afghanistan Investment and Reconstruction Task Force		Economic stabilization and infrastructure	Regional	Yes	No
DOC	Bureau of Economic Analysis	Hosted seminars for Iraqis that covered: basics of economic accounting, estimating methodologies, data dissemination, data revision policies.	Economic stabilization and infrastructure	Regional	Yes	No
DOC	National Institute of Standards and Technology	Assist Iraqi government in building a coherent system of standards and procedures for determining whether products meet requirements.	Economic stabilization and infrastructure	Regional	Yes	No
DOC	Census Bureau	Held workshops with experts from international community to provide technical assistance for the Iraqi census.	Economic stabilization and infrastructure	Regional	Yes	No
DOC—ITA	Iraq Investment and Reconstruction Task Force	Privatize state-owned enterprises, revamp food distribution, establish banking sector, eliminate corruption, facilitate housing construction and land ownership.	Economic stabilization and infrastructure	Regional	Yes	No
DOT—FHA	Baghdad Technology Transfer Center	Exchange transportation technology, provide technical assistance to Iraqi road builders, and establish an engineering training program for Iraqi engineers in the United States.	Establish or restore essential services	Regional	Yes	No

Table B.4 (continued)

Agency	Program Title	Program Objective	Primary MME	Geographic Focus	Indigenous Forces	Coalition Forces
DOC—NOAA	National Geodetic Survey	Collaborated with the U.S. Army to build the Iraqi Geospatial Reference System.	Establish or restore essential services	Regional	Yes	No
DOC—NTIA	Office of Spectrum Management	Assisted in the development of an independent communications regulatory commission in Iraq.	Establish or restore essential services	Regional	Yes	No
DOC	Bureau of Industry and Security	Assisted defense conversion in Iraq, expedited the delivery of materials to produce lightweight body armor, and supported emergency preparedness and critical infrastructure protection requirements (with cooperation from the Department of Homeland Security).	Economic stabilization and infra-structure	Regional	Yes	No
DOL	Bureau of International Labor Affairs	Provide training, job counseling, job finding, and institutional training to promote the reintegration of demobilized combatants.	Economic stabilization and infra-structure	Regional	Yes	No

Table B.5
Major Allies' BPC for Stability Operations Programs

Country	Program Title	Primary MME	Geographic Focus	Focus Countries	Indigenous Forces	Coalition Forces	Task Description
Australia	Regional Assistance Mission to Solomon Islands	Establish and maintain a safe and secure environment; Representative and effective government; Justice and reconciliation	Regional	Southwest Pacific	Yes	No	Provide protective services; conduct civilian police operations; provide interim public order
Australia	Australian-Indonesian Agreement on Framework for Security Cooperation	Safe/secure environment	Regional	Southeast Asia	Yes	No	Broad agreement to coordinate security forces and intelligence support
Canada	Directorate of Military Training Cooperation Programme	Safe/secure environment	Global	Newly independent, non-NATO commonwealth countries	Yes	Yes	Provide language training, professional development/staff training, and peace support operations training
France	Reinforcement of African Capabilities to Maintain Peace	Establish and maintain a safe and secure environment	Regional	Africa	Yes	Yes	Deliver and maintain military used equipment and provide peacekeeping training to friendly African nations

Table B.5 (continued)

Country	Program Title	Primary MME	Geographic Focus	Focus Countries	Indigenous Forces	Coalition Forces	Task Description
Germany	Bundeswehr's U.N. Training Center	Safe/secure environment	Global	U.N. member countries	Yes	Yes	Provide preparation of military operational contingents and regular training for international military observers as well as for civilian personnel
Germany	Kofi Annan Int. Peacekeeping Training Centre	Safe/secure environment	Regional	Africa	Yes	Yes	Develop regional orgs' capacity to conduct peacekeeping training
Turkey	Turkish General Staff Partnership for Peace Training Center	Establish and maintain a safe and secure environment	Global	Eurasia	Yes	Yes	Provide education and training for international security forces
Turkey	Center for Excellence Against Terrorism	Establish and maintain a safe and secure environment	Global	Eurasia	Yes	Yes	Provide counterterrorism education and training for international security forces
United Kingdom	British Military Advisory and Training Team	Establish and maintain a safe and secure environment	Global	West Africa, Eastern Europe	Yes	Yes	Send permanent forces to develop general needs of armed services; long-term focused
United Kingdom	British Peace Support Team	Establish and maintain a safe and secure environment	Regional	South Africa, East Africa	Yes	Yes	Send specialized forces to address specific needs of armed services; near-term focused
United Kingdom	Exercise Green Eagle	Economic stabilization and infrastructure	Regional	West Africa	Yes	No	Provide economic stabilization and infrastructure

Generic Indicators for Case Studies

This appendix provides a list of the generic output and outcome indicators that were used as the basis for measuring the effectiveness of various types of security cooperation tools (or "ways") in meeting the stability operations objective of "establishing a safe and secure environment" in the assessment case studies described in Chapter Four and Appendix D.

Conferences/Workshops/ Info Exchanges		Defense-Mil Contacts		Education	
Output	Outcome	Output	Outcome	Output	Outcome
Number participants per country		Number contacts/ year		Number countries and number of courses per year	
Appropriate representation (i.e., Interagency/ NGOs)	Best practices communi- cated	Number sta- bility opera- tions-related events		Number billets/ country	Stability op- erations built into national plans
Percent participants presenting	Repeat atten- dance of orga- nizations	Appropriate representa- tion sent	Assistance provided as result of contact	Appropriate representa- tion (i.e., Interagency, NGOs)	Relation- ships maintained
Agenda on safety and security	Tested new ideas during meetings	Agreement reached on building safety and security capacity	Agreement to employ or deploy capability in stability operations	Length of course	Institutional- ized goals into strategy, planning, training and doctrine
Number standing subgroups	New/existing formal/infor- mal networks leveraged to support safety security processes		Agreement resourced and executed	Quality of course as- sessment	
Follow-on effort determined					
Better under- standing and appreciation for other groups/ agencies	Provided mechanisms to discuss safety/ security issues				
Consensus around safety and security issues	Increase in conference participant interactions				
Dissemination of participant con- tact information and conference proceedings					

Training		Exercises		Equipment/ Infrastructure	
Output	Outcome	Output	Outcome	Output	Outcome
Number billets per country		Number countries/ participants and number of exercises held in a year		Appropriate amount/ type equipment/ infrastructure	
Skills acquired	Quality of trained forces as demonstrated in stability operations	Exercised existing capabilities	Adoption of common standards and concepts of operation	Necessary training provided to maintain equipment	Effective operational use of equipment
Level of interoperability (standards)	Capabilities deployed or employed	Operational/ technical problems identified	Resolved operational problems	Sustainment plan in place	Equipment sustained over time
Number soldiers/units trained per year	Training institutionalized in country	Appropriate representation sent	Successful deployment of units in support of stability operations	Logistics package of support in place	Equipment incorporated into logistics plan
Number trainers trained per year					
Appropriate representation					

Case Studies

The study team selected three case studies from Chapter Four to illustrate, in greater depth, the assessment process. We include two regional/coalition cases—Civil Military Emergency Preparedness (CMEP) program and the DoD regional centers—and one indigenous case—African Contingency Operations Training and Assistance (ACOTA)—to illustrate both types of partner capacity building approaches. We chose our most data-rich cases where partner and U.S. officials were consulted. Moreover, each of the cases represents different mixes of security cooperation ways. For example, CMEP exemplifies exercises, the DoD regional centers exemplify education/workshops, and ACOTA exemplifies the way of training.

Case Study 1: Civil-Military Emergency Preparedness Program

The U.S. Army's International Civil-Military Emergency Preparedness program encourages transboundary cooperation on emergency preparedness among countries that participate in NATO's Partnership for Peace program through joint disaster preparedness exercises. CMEP provides a forum for information exchange through the CMEP Council, which is especially critical since the national emergency response plans in the region are all classified.[1] In addition, the relatively nascent

---

[1] The CMEP Council for South Eastern Europe is an independent multinational structure, which was formed by CMEP initiative in 1999–2000 and includes 10 partner countries in

form of many of these states' disaster preparedness systems makes them ripe for international collaboration and discussion to identify best practices that can improve countries' indigenous disaster response. CMEP is overseen by the Office of the Secretary of Defense/Partnership Strategy and Headquarters, Department of the Army Stability Operations Division (G-35 SSO), and implemented by the U.S. Army Corps of Engineers. It is the sole U.S. Army-controlled program that has a stability operations application for building partner capacity. CMEP contributes to the end state of "safe and secure environment" through the way of exercises.

The study team conducted research on CMEP table top exercises (TTXs) through a review of relevant strategy documents, briefings, and focused discussions with U.S. officials and program managers from OSD/Partnership Strategy, U.S. Army Corps of Engineers, G-35 SSO, and with partner country officials. Additionally, in order to gain a deeper understanding of the CMEP program and its application vis-à-vis partner capacity building, the study team participated in a week-long CMEP TTX entitled ALBATROSS 07 in Batumi, Georgia in February 2007. The event was conducted within the framework of EUCOM's Black Sea Initiative.[2]

In particular, the team conducted an in-depth study of Romania's civil-military preparedness program and examined how, if it all, it has been impacted by CMEP. We selected Romania because Romanian officials have participated in the CMEP framework since the program's inception in 1995. For our analysis, the team spoke with representatives from the Ministry of General Inspectorate for Emergency Situ-

---

southeast Europe and western Eurasia. Italy and Greece are observers. The CMEP Council facilitates learning about other countries' processes, cooperative planning for response to disasters, understanding and cooperative use of procedures and population of databases of the Euro-Atlantic Disaster Response Coordination Center, and cooperation in planning and conducting civil protection exercises under civilian leadership (mostly ministries of interior) with military support.

[2]   The Black Sea Initiative was developed in close cooperation with EUCOM, and is one of four components ("pillars") of EUCOM's Black Sea Strategy, the others including Airspace Awareness, Border Security, and Maritime Domain Awareness.

ations (GIES) in the Ministry of Interior in Romania who have long participated in CMEP events.

The scenario for the Batumi TTX included a terrorist incident in which a mass oil spill resulted off the coast of Batumi, Georgia in the Black Sea. The U.S. Army Corps of Engineers provided Geospatial Information System (GIS) engineers and cartographers to improve the capacity of the partners in working with geospatial data in a crisis. During the TTX we had an opportunity to discuss CMEP's impact with several partner countries in southeast Europe and Eurasia. Participants included representatives from the respective ministries of emergency preparedness and disaster response from Georgia, Azerbaijan, Bulgaria, Moldova, and Romania. The following countries sent observers: Turkey, Ukraine, Armenia, and Croatia. Several international organization representatives also attended to facilitate the discussions and training: U.N. Office for the Coordination of Humanitarian Affairs and NATO's Euro-Atlantic Disaster Response Coordination Centre. Several NGOs also attended.[3]

Inputs

Money. CMEP is funded by the Warsaw Initiative Fund, overseen by OSD/Partnership Strategy. The CMEP budget has steadily decreased, from $3.1 million in 2005, to $2.3 million in 2006, to $1.2 million in 2007. The average cost of a CMEP TTX is $350,000–$400,000.

Manpower. According to CMEP program managers, the average number of U.S. officials taking part in CMEP events is ten, depending on the phase of the TTX process (fewer for the planning events and more for the actual execution).

Outputs

Quantity of exercises held. In prior years, CMEP averaged about four TTXs per year. However, with the decrease in funding, CMEP typically facilitates one major multinational TTX per year in a CMEP partner nation.

---

[3] International Red Cross/Red Crescent, the Organization for the Prohibition of Chemical Weapons, and Japan's International Regional Center for Disaster Management.

Operational and technical problems identified. During the CMEP TOMIS exercise in Romania in 2005, multiple problems were identified with the new structure, including the need to

- Increase training of intervention forces.
- Make improvements to the national insurance system.
- Coordinate national response plans with international organizations.
- Increase public information/education and prevention with regard to preparing for disasters.

First, the Romanians identified problems with the training of the intervention forces. They determined that these forces need to be available for a wide range of responses, not just fire responses. Second, the TOMIS TTX also highlighted the fact that Romania did not have an insurance system in place for earthquakes and floods. Third, TOMIS also highlighted the need for Romania to coordinate its response standards with those of the U.N. and the World Health Organization. Finally, CMEP TTXs provided Romania (as well as other nations) with ideas to help improve their public information and education systems. CMEP has recently begun to focus on this issue. According to Romanian officials, the specific focus should be on public information/education and prevention with regard to dealing with natural and man-made disasters. Public information was heavily included in the scenario for the Batumi TTX.

Appropriate representation sent. Based on the study team's observations and discussion with participants, Romania consistently sends the appropriate individuals to CMEP events. During the Batumi TTX, the Romanian delegation openly shared lessons from the reform of their emergency response system. They also took on a regional leadership role with the other participants in the various functional cells of the TTX.

Capabilities successfully exercised. Since 1994, CMEP has assisted many partners in hosting TTXs for emergency planning on a regional basis.[4] Every member nation of Partnership for Peace has participated in at least one CMEP-assisted TTX. Specifically, Romania tested and validated a new civil defense structure and exercised existing GIS capabilities at the TOMIS TTX in 2005. CMEP TTXs use NATO Euro-Atlantic Disaster Response Coordination Centre procedures as a tool for building procedural interoperability.

Outcomes

Adoption of common standards. Romania's national response plan deals mostly with coordination of the intervention. Therefore, Romania tested the response capabilities of their new response plan in conjunction with the CMEP TOMIS exercise in September 2005, and this system was apparently used in response to an outbreak of bird flu the following year. According the U.S. and Romanian officials, CMEP also greatly helps to facilitate Romania's interagency coordination. Moreover, Romania has taken on a supporting role with other CMEP members, such as Croatia. Above all, CMEP procedures led to the creation of a network of networks among the Black Sea countries, to which Romania is firmly connected.

Operational and technical problems resolved. Romania first learned about the specifics of the U.S. National Response Plan at the TOMIS TTX, then eventually began to realign its civil defense structure in the defense ministry to become the GIES in 2005. Next, Romania created its own plan for risk mitigation in 2006 based on the National Response Plan as a model. Subsequently, Romania developed a U.S.-influenced perspective of civil-military and transboundary cooperation for disaster preparedness and response. However, Romania's new plan is classified and Romania is still in the development stage of its national emergency response system.

As a result of the TOMIS TTX, Romania sought and obtained a loan from the U.N. to enhance its emergency management capabilities

---

[4] These include Albania, Armenia, Azerbaijan, Belarus, Bosnia-Herzegovina, Bulgaria, Croatia, Czech Republic, Estonia, Georgia, Hungary, Kazakhstan, Kyrgyzstan, Latvia, Lithuania, Macedonia, Poland, Romania, Serbia, Slovakia, Tajikistan, and Ukraine—with current planning for an event to be hosted by Moldova.

(i.e., training its forces) to improve earthquake response capabilities. The U.N. loan helped the GIES create an information management system focused on a public awareness campaign and an education curriculum. The GIES is currently working to create such a system. This topic was discussed extensively during the Batumi TTX.

Also, during the Batumi TTX, Romania proposed that an independent source be invited to assess the civil emergency preparedness systems of all Black Sea Initiative countries.[5] A follow-on conference hosted by the Black Sea Forum in Romania revisited this topic in July 2007. Since that event, a contract was awarded to a U.S.-based government think tank to conduct civil emergency preparedness assessments in many, but not all, of the Black Sea littoral countries. CMEP encourages partners to look at civil emergency planning and seriously assess where they are in terms of capabilities and shortfalls.

External Factors
Process factors. New procedures have been adopted in the Romanian GIES because of ideas they were exposed to at CMEP TTXs.

Additional Security Cooperation Ways. Mutually supporting regional initiatives such as Stability Pact, NATO Euro-Atlantic Disaster Response Coordination Centre and Civil Emergency Planning Directorate, and Black Sea Initiative enable CMEP's success. Romania supports key CMEP concepts because they are reinforced by many regional capacity-building organizations.

Country Factors. Although Romania remains an enthusiastic CMEP partner, economic conditions are an inhibitor to implementing its robust reform agenda. Romania still suffers from widespread poverty and corruption. Public debt is 21 percent of gross domestic product.

Overall Assessment
Participation in regional stability operations arrangements. CMEP's main strength appears to be its ability to build relations, or

---

[5] Phase I would be a vulnerability and needs assessment; Phase II is an implementation plan for upgrading systems.

a "network of networks of people" in a regional and multilateral context. Further, CMEP works through existing regional organizations, and has an ability to engage civilian government agencies such as the emergency management ministries, border guards, and ministries of interior that other military-to-military programs cannot. More formalized alumni networks may be useful for CMEP, particularly as the program's greatest benefit appears to be building ties on a multilateral level to increase confidence, trust and interoperability.

CMEP program managers might consider encouraging partners to either declassify parts of their national response plans that pertain to regional cooperation so that these plans could be shared. In such a case, CMEP TTXs could be used to exercise actual national response plan on a regional basis.

CMEP could also potentially be used a distance learning tool for stability operations through the sharing of GIS map data. However, it is not utilized in this way currently.

In terms of coordination with other, similar U.S. programs, CMEP has had mixed success. For example, coordination with the State Partnership Program has not always been apparent but is improving.[6] In addition, CMEP has not formally coordinated events with the Defense Threat Reduction Agency's International Counterproliferation Program, for example, despite the fact that weapons of mass destruction scenarios are used in the TTXs and field training exercises of each respective program.

Overall, in Romania CMEP is achieving the goal of a safe and secure environment. New procedures have been adopted, and lessons from others have been taken into account. However, Romania's poor economic state is inhibiting the implementation of CMEP's robust reform objectives.

---

[6] For example, the CMEP New Hampshire program was developed as a means of filling a gap for conceptual and pragmatic understanding of U.S. state and local emergency preparedness and disaster response capabilities, where SPP was not able to host target countries as desired by defense policy considerations.

Case Study 2: DoD Regional Centers Stability Operations Courses and Conferences

The Department of Defense operates five regional centers that (in part) provide educational courses designed to build partner countries' capacity to contribute to stability operations. Four of the centers have courses related to stability:

- George C. Marshall Center (GCMC) in Garmisch, Germany.
- Asia-Pacific Center for Strategic Studies (APCSS) in Honolulu, Hawaii.
- Near East South Asia Center for Strategic Studies (NESA) at Fort McNair, National Defense University (NDU), in Washington, D.C.
- Center for Hemispheric Defense Studies (CHDS) at Fort McNair, NDU.

The fifth center, the Africa Center for Strategic Studies at Fort McNair, NDU, does not offer stability operations courses, but focuses instead on general leadership programs.

The centers contribute to the goal of a "safe and secure environment" through the "ways" of education and conferences in the respective regions.

The RAND Arroyo Center team analyzed information about all four of these centers, and also studied the NATO School in Oberammergau, Germany, which teaches stability operations-related courses at the operational and tactical level.[7] Research for this case study was conducted through a review of programs of instruction, relevant briefings, course assessments, where possible, and focused discussions with the regional center directors, deans, and course instructors. Where possible, feedback from the partner countries was included (i.e., if survey data were available).

---

[7] The NATO School receives some funding from U.S. sources, including International Military Education and Training, for their courses.

We considered how each of the regional centers approach stability operations within their curricula. We started with the rather vague context and guidance provided by OSD to the regional centers. Indeed, OSD does not tell the centers which capabilities to develop in which partners in a stability operations context, and the OSD guidance does not provide such a menu of programmatic or resourcing options.

To varying degrees, the regional centers are in the process of developing new course curricula or modifying existing courses as best they can to reflect the latest stability operations guidance, as discussed in Chapter Two. The centers are operating in diverse regions, and thus their terminology differs in the context of stability operations. For example, in the EUCOM AOR, it is more appropriate to use "peace-keeping" as the preferred term. In PACOM, "coalition" is not a popular term due to regional animosities. The centers are free to develop a curriculum that reflects these cultural sensitivities.

Overall, we found that all of the courses taught differ in terms of scope, target audience, and duration. There does not appear to be any overarching strategy or guiding principles on the educational curriculum. In addition, there appears to be very little collaboration on the topic of stability operations among the regional centers. Much can be gained from such a collaboration—not just on the stability operations-related courses, but other functional courses as well.

Because the centers are developing their own, independent courses with little interaction among course leaders and deans on the subject of stability operations, they may be missing opportunities to collaborate and share best practices. For example, the stability operations-related courses taught at the centers could benefit from the inclusion of experts from a variety of disciplines and agencies. The problem is that only government officials are eligible, by law, to attend the courses (with expenses paid by the centers).[8]

However, CHDS and APCSS both reported that they are in the process of developing a distance-learning tool for stability operations

---

[8] Military and civilian officials from the ministries of defense, foreign affairs, emergency situations, justice, interior, etc. are eligible, but not NGOs or academics, for example.

courses. Collaboration among the instructors could lead to resource savings and improvements to the approach.

Another interesting aspect to the guidance is that it does not necessarily come with additional resources to develop such a course. This has been a problem for some centers. Here again, greater collaboration between them could facilitate the sharing of innovative funding ideas. For example, APCSS funded its new core stability operations course out of existing resources—it reduced its 12-week core security cooperation course to six weeks. The proceeds from the shortened security cooperation courses funded the new three-week stability operations course. Moreover, CHDS reduced its main survey course from 72 to 56 students to accommodate an additional 16 students in its new course on advanced stability operations, which commenced in June 2007.

Table D.1 compares some of the key commonalities and differences in four of the regional centers in terms of their respective stability operations or related courses.

For our in-depth analysis, the study team focused on the George C. Marshall Center and in particular on Romania's participation in stability operations-related courses and regional center conferences. Located in Garmisch, Germany, GCMC's mission is to create a more stable security environment by advancing democratic defense institutions and relationships; promoting active, peaceful engagement; and enhancing enduring partnerships among the nations of America, Europe, and Eurasia. This is accomplished through tailored advanced professional education and training of military and civilian officials and by applied research.

GCMC's stability operations course, titled *Program for Peace Support and Stability Operations*, commenced in 2004 and remains an elective. It is taught three times per year and lasts for three weeks. The course consists of four thematic modules featuring presentations by expert U.S. and international civilian, military and governmental practitioners. Each module is followed by small group seminars, in which participants debate and exchange ideas on the issues presented.[9] In

---

[9] Module I: General Peacekeeping; Module II: Security and Stability; Module III: Transition and Reconstruction; Module IV: Capacity Building.

Table D.1
Comparison of Regional Centers' Approach to Stability Operations

Commonalities	Differences
All want to include stability operations as outreach in addition to core/elective courses.	Some courses held annually (e.g., CHDS), others are multiple times per year (e.g., GCMC, APCSS).
All have an interest in helping partners to develop new doctrine for stability operations.	Some advocated for additional resources to create new stability operations course (e.g., GCMC), while others reduced length of core course to fund new stability operations course (e.g., APCSS).
All have difficulty resourcing stability operations as a core course (NESA opted to not create a course).	Target audiences differ; some target only alumni (e.g., CHDS).
All have difficulty funding non-governmental participants. Some workarounds are created (e.g., APCSS organized a workshop on stability operations for NGOs that overlapped with a stability operations course).	Some include U.S. interagency participants; some do not.
	All are moving at different speeds in course development.
Several centers are interested in creating a distance learning course about stability operations.	Outreach approach differs; some target partners deploying to Iraq and Afghanistan (e.g., GCMC); others hold conferences to discuss stability operations in a more conceptual way.
	All assess the value of stability operations courses differently.

addition, several case studies, exercises, and an extended field trip serve to reinforce the topics discussed.[10]

According to the program of instruction, the purpose of this course is to provide mid-level military officers, police administrators, and civilian managers with the necessary theoretical and practical knowledge required to build national capacities to conduct peace support and stability operations in post-conflict and crisis areas. Besides serving to develop a common understanding of the subject matter, this course showcases capacity-building resources, facilitates contact between participants for better integration into the stability operations community, and enables individual nations to cooperate successfully on future missions.

---

[10] Discussions with course leaders, February and May 2007, Garmisch, Germany.

Inputs

Money. GCMC relies on Warsaw Initiative Funds and the Counterterrorism Fellowship Programs to support their students' attendance. No additional resources have been provided for the development of a new core stability operations course.

Manpower. Two instructors team up to teach this course, with several subject matter experts serving as guest lecturers in their specific areas of expertise.

Outputs

Quantity of events per year. For GCMC courses, students can choose three subjects from a host of electives on security studies topics. A total of six Romanians attended the five courses that were held between 2004 and mid-2007.

Romania has hosted numerous GCMC conferences since 2000, including two on stability operations-related topics.

Appropriate representation. For GCMC courses, representatives from the ministries of Defense, Interior, and Foreign Affairs regularly attend GCMC courses every year.

For GCMC conferences, Romania sends appropriate representatives to events in relatively large numbers and occasionally funds the travel of its participants. Attendance varies according to topic, but in general government officials and NGOs from a variety of sectors regularly attend GCMC conferences. Romania also provided five key speakers to GCMC conferences from 2005 to 2007.[11]

Outcomes

Forthcoming with new ideas during meetings. For GCMC courses, this data is not available because GCMC encourages open discussions and guarantees confidentiality to encourage freedom of opinion on critical issues. Therefore, GCMC declined to provide information on participation and performance so as not to violate that trust with course participants.

---

[11] Based on Arroyo's observations during six separate conferences held in Croatia, Germany, Macedonia, Bulgaria, Lithuania, and Montenegro from 2005 to 2007, as well as discussions with GCMC and partner country representatives.

For GCMC conferences, Romania's active involvement in conferences in southeast Europe has facilitated regional cooperation and the sharing of lessons learned and best practices, for example, in security sector reform and emergency preparedness.

Relationships maintained/significant follow-on events occur. For GCMC courses, it appears that Romania's alumni network is one of the most dynamic among all of the partners. This is evidenced by Romania's ability to both organize and fund conferences on key strategic issues, as well as to stay firmly connected to the Marshall Center. For example, the Romanian alumni program organized a high-level conference in October 2006 on energy security that featured the president of Romania and was attended by other high-level officials. The conference was highlighted by often heated but largely productive dialogue between policymakers in the region and Russia's Gazprom.

For GCMC conferences, Romanian officials are great networkers, enthusiastically sharing their thoughts on best practices with other participants. They often volunteer to take the workshop leadership role. However, there is no clear framework to facilitate follow-up on various issues raised during the conferences. Conference participants are not included in the alumni networks in the same way as are Marshall Center course participants.

External Factors

Process factors. As mentioned previously, the regional centers, including the GCMC, are largely independent in their efforts to design their own stability operations courses. While this approach promotes their academic freedom and integrity, they may be missing opportunities to collaborate and share best practices. The various courses could benefit from the inclusion of experts from a variety of disciplines and agencies.

As mentioned above, legislation will not allow for funding for NGOs and academics to take part in stability operations courses, which usually means those experts are excluded. However, work-arounds have developed. For example, APCSS organized a three-day workshop on stability operations for which NGO representatives were eligible to receive funding to attend. The dates of the workshop aligned with the three-week stability operations course in Honolulu. This creativity allowed for NGO representatives to contribute to the stability operations course, using the appropriate funding sources.

Moreover, it appears that the coordination between GCMC and the NATO School on the topic of building partner capacity for stability operations is limited. Traditionally, GCMC's focus has been at the strategic level, whereas the NATO School is at the operational level. However, some of the Marshall Center's outreach activities, such as preparing partners for deployment to Afghanistan, include discussions on operational-level issues. Romania takes part in both GCMC and NATO School courses.

Additional security cooperation ways. There is some coordination on a limited basis with the State Partnership Program, though not in Romania. Other GCMC outreach activities are focused on partner countries that are preparing to deploy to Afghanistan, such as those in Central Asia, the Caucasus, and the Baltics.

Country factors. While Romania remains an enthusiastic GCMC partner, economic conditions can be an inhibitor to implementing its reform agenda. Romania still suffers from widespread poverty and corruption. Public debt is 21 percent of gross domestic product.

Overall Assessment

Participation in regional stability operations arrangements. Romania is a net contributor to stability operations arrangements, for example, in Afghanistan, Iraq, and the Balkans. Romanian officials openly share operational lessons with other regional partners during GCMC events. However, it would be a stretch to claim that Romania participates in stability operations arrangements *because* of its ties to the Marshall Center.

Case Study 3: Africa Contingency Operations Training Assistance Program

The ACOTA program provides training to African militaries in order to increase partner countries' capacity to conduct peace support operations. Based on Arroyo's assessment framework, ACOTA uses the way of training to build indigenous partner capacity toward achieve the end of a "safe and secure environment" in Africa. Key ACOTA goals include imparting peace support operation skills for troops and for battalion-level command and staff, and building and sustaining partner countries' capacity to train their own peace support forces.

While African troops occasionally deploy outside of Africa,[12] ACOTA's primary focus is to address the need for peace support on the African continent: two-thirds of African nations experienced a civil war between 1991 and 2005.[13] There have been 19 U.N. peacekeeping missions in Africa since 1960,[14] six of which are ongoing.[15] Given DoD's increasing investment in Africa with the creation of the U.S. Africa Command (AFRICOM),[16] it is critical to understand existing U.S. efforts to enhance the stability capabilities of African allies and the obstacles those efforts face.

ACOTA falls under the auspices of the Global Peace Operations Initiative (GPOI), a G-8 initiative.[17] ACOTA is the George W. Bush administration's successor program to the Clinton administration's African Contingency Response Initiative. In contrast to the African

---

[12] For example, as of June 2007, 868 Ghanian troops and 77 Tanzanian troops were deployed with the U.N. Interim Force in Lebanon.

[13] Our study relied on Ted Robert Gurr, Monty G. Marshall, and Keith Jaggers, "Polity IV Database, 1800–2004." For a more recent version of the database, see Monty G. Marshall and Keith Jaggers, "Polity IV Project: Political Regime Characteristics and Transitions, 1800–2008." As of February 2010:
http://www.systemicpeace.org/polity/polity4.htm

[14] In accordance with AFRICOM's focus on all countries on the African continent with the exception of Egypt, this count does not include U.N. missions focused on stabilizing relations between Egypt and Israel.

[15] Regional bodies such as the African Union and the Economic Community of West African States have also led numerous African peacekeeping missions, all of which eventually became U.N. missions.

[16] For background on the genesis of and plans for AFRICOM, see the AFRICOM transition team web site, and Lauren Ploch, "Africa Command: U.S. Strategic Interests and the Role of the U.S. Military in Africa," CRS report RL34003, October 2, 2009.

[17] For more information on GPOI, see Nina Serafino, "The Global Peace Operations Initiative: Background and Issues for Congress," CRS Report RL32772, updated March 19, 2009.

Contingency Response Initiative's emphasis on training in nonlethal peacekeeping skills in accordance with U.N. Charter Chapter VI, ACOTA entails Chapter VII training of lethal peace enforcement techniques, reflecting lessons learned from the high-threat environments that peace support operations faced in the African wars of the early 1990s, such as Sierra Leone. ACOTA also added an emphasis on capability sustainment through training the trainer.

The Bureau of African Affairs at the U.S. Department of State executes the ACOTA program in collaboration with the Africa Section of the OSD's International Security Affairs office. These offices also coordinate closely with State's Bureau of Political-Military Affairs, which oversees GPOI and manages ACOTA funding through the Foreign Operations Appropriations Peacekeeping Account. An interagency Policy Development Oversight Committee, with leadership from State, OSD, the Joint Staff, and the National Security Council, provides high-level direction for ACOTA, including selecting new countries to invite into the program and evaluating possible suspensions of partnership when partner militaries conduct activities that conflict with U.S. policy interests.[18]

In order to describe and assess the ACOTA program, the Arroyo study team collected broad programmatic information on ACOTA along with detailed information on deployment outcomes for four ACOTA partner countries: Botswana, Nigeria, Rwanda, and Senegal. The latter three countries all contributed troops to the African Union Mission in Sudan (AMIS), providing an opportunity to study ACOTA training effectiveness in an operational environment.[19] The team reviewed U.S. government documents, U.N. documents, and scholarly articles, and spoke with U.S. and foreign officials. At the State Department, the team spoke with officials from the ACOTA Program Office and the Bureau of African Affairs, as well as officials from the Bureau

---

[18] For example, the United States suspended ACOTA training to Uganda following Uganda's invasion of the Democratic Republic of the Congo in 1998. The U.S. resumed the partnership in 2007.

[19] Botswana provided airlift support for Rwandan troops serving with AMIS, but has not otherwise contributed to peacekeeping missions in Africa since ACOTA was established in 2002.

of Political and Military Affairs. At DoD, the team interviewed officials from OSD, EUCOM (before AFRICOM was established), and U.S. Army Europe, along with former defense attachés who served in the Democratic Republic of Congo, Liberia, Nigeria, and Sierra Leone. The team also spoke with foreign military officials from Benin, Nigeria, Mali, Senegal, and Rwanda. We also held discussions with experts from the Africa Center for Strategic Studies, the U.S. Institute of Peace, the U.S. Army War College, and relevant private-sector contractors. Additionally, the team spoke with contractors for the U.S. government who are stationed in military observer positions with the African Union peacekeeping force in Darfur.

Inputs

Money. The average ACOTA investment per soldier trained and equipped entirely by the United States is roughly $3,700.[20] The per-course cost of training depends on the partner country's progress toward achieving the ability to teach the training courses themselves. If a training course is run entirely by U.S. trainers, then the cost of training one battalion is $1.2 million. If the U.S. contribution to training is roughly 50 percent, then the cost to the United States is roughly $650,000. If the partner country has reached the full ability to instruct the course themselves, then the U.S. role will be minimal, with only two to three U.S. trainers to observe and advise the host country's trainers when appropriate, and the U.S. cost will be approximately $150,000.[21]

In addition, ACOTA spends about $1 million on equipment for each battalion trained, depending on an assessment of need. ACOTA provides only the nonlethal equipment that enables trainers and trainees to participate fully in the courses, such as white boards, uniforms, sleeping bags, tents, boots, generators, radios, first aid, and water puri-

---

[20] This figure is a RAND Arroyo Center calculation based on State Department estimates: DOS officials said that $1.2 million is a rough estimate of the cost to train one partner country battalion (in cases in which no partner trainers are contributing), and $1 million to equip one battalion. We assume a 600-person battalion.

[21] Information provided by officials in the DOS Bureau of African Affairs and the Bureau of Political and Military Affairs.

fication equipment. The State Department provides GPOI funding (separate from ACOTA) for lethal equipment on a case-by-case basis to African countries prior to deployment.[22] However, U.S. and African officials noted that deployed African battalions are often ill-equipped and could benefit from additional training using the lethal equipment they use on their missions.

As Figure D.1 illustrates, ACOTA funding has increased in recent years, reflecting greater U.S. investment in building partner capacity with the creation of GPOI in 2005. In addition, the number of partner countries also has increased.

Manpower. Until a partner country has attained the skills to independently train its military for peace support operations, the United States sends a team of roughly 15 to 20 contractor trainers to each course. Officials said that although it would be ideal if all ACOTA training was provided by uniformed servicemen, given their credibility with trainees, the significant U.S. deployments to Iraq and Afghanistan preclude additional uniformed participation in ACOTA. The presence of a small cadre of uniformed mentors along with contractors thus represents a reasonable solution to the manpower shortage. These highly qualified contractors are typically retired U.S. military enlisted and officers, along with three to six uniformed officer mentors.[23]

Capabilities. After conducting a pre-training assessment in consultation with the partner, ACOTA trainers seek to tailor each training course to a partner country's needs. A typical ACOTA training event consists of five weeks of command and staff training and four weeks of soldier skills. Staff training focuses on topics such as command and control of peace support operations and military decisionmaking processes. Troop trainings are focused on light infantry and small-unit tactics, such as conducting patrols and checkpoints. Both staff and troops also receive training in general topics like human rights, HIV/ AIDS, and rules of engagement.

---

[22] According to a senior OSD official, DoD does not provide equipment because it cannot do so unless the recipient country has signed an Article 98 agreement, which is a bilateral nonsurrender agreement of U.S. citizens to the International Criminal Court.

[23] This responsibility will soon be transferred to AFRICOM.

Figure D.1
ACOTA Funding and Number of Partner Countries

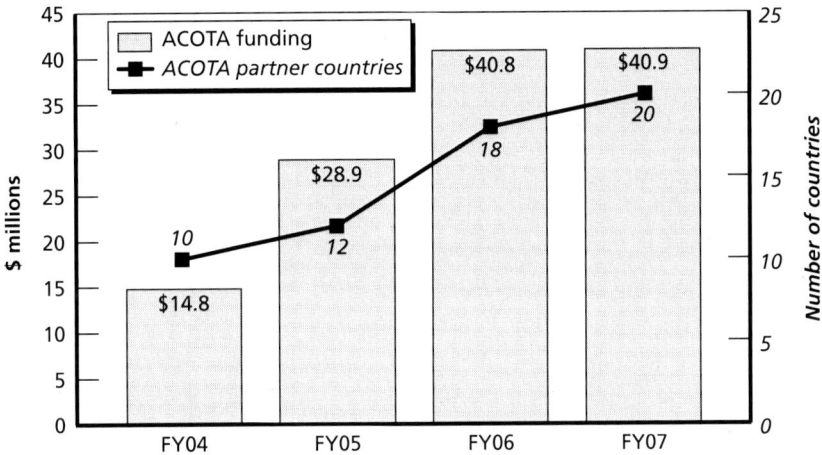

SOURCE: RAND Arroyo Center analysis of data provided by State Department Bureau of African Affairs.

RAND *MG942-D.1*

Additionally, the United States typically provides initial training to host country trainers over three iterations of the battalion-level training program:

• First iteration: the host trainers observe.
• Second iteration: the training responsibility is shared equally.
• Third iteration: the host country trains and the U.S. personnel observe.

Following this sequence, the host country often continues to invite a small number of U.S. trainers to observe trainings and to provide input where relevant.[24]

---

[24] This information comes from DOS documents and interviews, including an interview with an ACOTA trainer.

Outputs

Number of units trained per year. The number of troops trained under ACOTA increased under GPOI in 2005. While detailed data is not available prior to FY 2005, ACOTA program officials report that roughly 9,000 troops were trained in FY 2004, whereas roughly 15,000 troops were trained in FY 2006. Table D.2 lists all ACOTA partner countries, the year in which the United States initiated the partnership, and training that those partners received since FY 2005, as of July 2007.[25]

Number of trainers trained per year. ACOTA's train-the-trainer component appears to be achieving its goal of fostering self-sufficiency and capability sustainment among partner countries. The number of host country trainers trained by ACOTA increased from 275 in FY 2005 to 909 in FY 2006. As of June 2007, 442 trainers were trained by ACOTA in FY 2007.[26] According to DOS and DoD officials, trained trainers typically can teach their own courses independently, but for support and oversight purposes, the United States continues to send approximately three trainers to each course after a partner country has achieved the ability to conduct training independently.

There has been variation in the rates at which countries achieve the capability to train independently. For example, although Rwanda and Botswana achieved self-sufficiency after ACOTA trained three battalions in each country, Nigeria achieved self-sufficiency after five battalions were trained. In contrast, Senegal has had 15 battalions trained and has not yet reached self-sufficiency.[27]

Appropriate representation. The United States works with the partner country in advance of training events in an effort to ensure that high-quality soldiers attend the ACOTA training, but these efforts yield mixed results. Ultimately, partner countries make the final

---

[25] ACOTA has authority to provide training for regional, multinational forces, and did so in 2006 and 2007 for staff of the Economic Community of West African States forces. The ACOTA program office also plans to train the African Standby Force in advance of its 2010 rollout.

[26] Discussions with State Department officials.

[27] According to ACOTA trainers, Senegal was expected to achieve self-sufficiency in 2008.

Table D.2
Training Received by ACOTA Partners

ACOTA Partners	Year ACOTA or African Contingency Response Initiative Partnership Began	Number of Battalions Trained Since FY 2005 (staff, troops, and trainers unless otherwise specified)
Benin	1999	2
Botswana	2004	1 (staff only)
Burkina Faso	2006	2 (1 of which was trainers only)
Cameroon	2007	0
Gabon	2005	3 (1 of which was staff only)
Ghana	1998	5
Ethiopia	2003	0
Kenya	2000	1 (staff only)
Malawi	1998	1
Mali	1999	2
Mozambique	2004	3 (1 of which was staff and trainers only)
Namibia	2006	0
Niger	2006	Less than 1 (357 troops)
Nigeria	2005	5
Rwanda	2006	12
Senegal	1997	12
South Africa	2004	37 medical personnel and 1 brigade staff
Tanzania	2006	0
Uganda	1998	0
Zambia	2004	0

SOURCE: State Bureau of Political-Military Affairs data.

decision about who receives training. One exception is that the State Department screens each soldier slotted to receive training in an effort to ensure that none are HIV positive and none has been convicted of human rights violations.[28]

One area of concern to officials is that partners sometimes send "composite" battalions to ACOTA training—that is, small pieces of disparate battalions that are combined as a unit for the first time during

_____
[28] The State Department has the lead for human rights vetting under the Leahy Amendment provisions.

the ACOTA training. Officials are concerned that training composite battalions limits unit cohesiveness of the battalions that eventually deploy, as only some of the deploying soldiers have received ACOTA training, while most have not. The ACOTA-trained battalions also do not always deploy as a unit, particularly if significant time elapses between the training and the deployment.[29] One Senegalese official stated that Senegal often sends composite battalions to ACOTA training and on deployment because of security concerns in the regions where the battalions are based; they do not want to leave their barracks unattended. One U.S. official argued that sending composite battalions is the prerogative of the partner country, and thus beyond the control of the ACOTA program.[30]

Outcomes

Capability deployed. In FY 2006, 79 percent of all African battalions that deployed on peace support operations globally had received ACOTA training.[31] There has been cross-national variation in deployment rates of trained troops: all 12 of the Senegalese battalions trained by ACOTA since FY 2005 have deployed in peacekeeping missions in Côte d'Ivoire, the Democratic Republic of the Congo, Liberia and Sudan. Eleven of Rwanda's 12 battalions that were trained have deployed in Sudan. Three of Nigeria's five trained battalions have deployed in Sudan. In contrast, none of Botswana's three trained battalions has deployed. In response, the United States has halted ACOTA training of troops from Botswana, with the exception of refresher training.

Improved capability demonstrated in peacekeeping operations. Determining the causal impact of ACOTA training on units' performance in peacekeeping missions with a high degree of certainty is difficult because of data limitations and external factors. However, U.S.

---

[29] The State Department has only begun capturing reliable troop-level data that will enable the tracking of individual soldiers from training to deployment. As a result, one cannot conduct a historical analysis of the extent to which trained battalions have been divided before deployment.

[30] Discussions with OSD officials.

[31] State Department FY 2006 Performance and Accountability Report.

officials, African officials, and nongovernmental experts all cited the usefulness of the ACOTA program, and most indicated that they perceived the program to improve significantly the professionalism and skill level of the units that received training. One U.S. official noted that nontrained battalions were more likely to run prostitution rings or demand bribes at checkpoints. Another official argued that ACOTA improved units' skills in responding to ambushes.[32]

Decisively ascertaining whether ACOTA training improved a unit's ability to perform effectively during peace operations would require one to observe the counterfactual: would trained units have performed differently if they had not received ACOTA training? This question cannot be answered directly, but it can be crudely approximated by comparing the performance of two units—one ACOTA-trained and one not ACOTA-trained—that faced similar deployment environments and share important background traits, such as country of origin. Such matched pairs of units are rare, given that in most recent deployment scenarios in Africa, all cases of multiple units deployed from one country were all ACOTA trained. However, in peace missions in Sierra Leone and Sudan, U.S.-trained Nigerian battalions rotated into the mission, replacing other Nigerian battalions that were not U.S. trained.[33]

Many U.S. observers who have served in Darfur, including former U.S. defense attachés who served in West Africa during the training of forces that deployed to Sierra Leone, State Department officials, and U.S. contracted military observers, all noted significant improvement in professionalism and skill level in the U.S.-trained Nigerian battalions compared to the nontrained Nigerian battalions that preceded them.[34] For example, a U.S. military observer to AMIS reported that

---

[32] Discussions with OSD officials.

[33] In Sierra Leone, the battalions were trained by an ACOTA predecessor program known as Operation Focus Relief, which trained seven Nigerian battalions during 2000 and 2001 in Chapter VII peace enforcement skills for the purpose of deployment with the U.N. mission in Sierra Leone. These battalions also later deployed in the U.N. mission in Liberia.

[34] U.N. reporting, which included surveying the civilian population, also praised the efforts of the United Nations Mission in Sierra Leone.

non-ACOTA-trained battalions would not take responsibilities such as guarding camp at night seriously. While this observer argued that the ACOTA-trained Nigerian troops he observed still fell far short of his expectations for effective peacekeeping and were not as skilled as Senegalese or particularly Rwandan troops, he found that they behaved more professionally than non-ACOTA-trained troops. For example, unlike the non-ACOTA-trained troops, they were serious about their responsibility to guard the camp all night. ACOTA-trained troops were also more likely to comply with rules of engagement when confronted with hostile forces, he said.

While these peace missions, particularly AMIS, have not achieved the end of a stable and secure environment, this analysis constitutes fairly strong support for ACOTA's positive impact on operational outcomes. That is, even if the peace support missions failed in some respects, it stands to reason that outcomes might have been worse in the absence of ACOTA training.

External Factors

Process factors. ACOTA assessments are also challenging due to the State Department's inconsistent methods for tracking reliable ACOTA data. In its data-collection efforts, the Arroyo team often received inconsistent figures on ACOTA accomplishments from two different State Department offices. In order to systematically collect reliable data, all offices involved in ACOTA should develop agreed-upon metrics that the U.S. government can use to track levels of skill attainment before ACOTA training as well as over time after training. Information on troop effectiveness during deployment outcomes would be particularly informative.

Country factors. Country-level factors that are external to the ACOTA program likely have substantial influence on whether ACOTA-trained forces are able to achieve stability when they deploy in peacekeeping operations. It is difficult to disentangle the influence of external factors from that of ACOTA-related factors, but the potential impact of external factors on the outcome is quite large. In all cases cited below, U.S. officials suggested these external factors as important influences on the respective countries.

With respect to Botswana's hesitancy to deploy troops on peace-keeping missions, State and OSD officials cited the country's wealth relative to other African countries as a crucial factor. Botswana's gross domestic product (GDP) per capita in FY 2006 was $15,020, which was more than five times that of Nigeria ($1,230), Rwanda ($1,430), and Senegal ($2,270).[35] As a result, Botswana does not have the same economic motive to participate in peacekeeping operations as do many African countries. The U.N. pays roughly $1,000 per month per troop deployed, a portion of which goes to the government from which the troops originate.

With respect to Nigeria, U.S. observers to the African Union mission in Darfur cited a generally low level of military capacity compared to Rwandan and Senegalese troops serving there, and largely attributed this difference to historical factors that have eroded Nigerian military institutions. In particular, a history of hostile relations between Nigerian civilian governments and national militaries often looms large. Since achieving independence in 1960, Nigeria has endured seven military coup attempts: five successful coups in 1966, 1975, 1983, 1985, and 1993; and two failed coups in 1976 and 1990.[36] In contrast, Senegal has not experienced any coups since independence in 1960. These tumultuous civil-military relations in Nigeria may have eroded governmental and civilian support for military institutions, accounting in part for Nigeria's relatively low baseline military capacity despite its regional power and size. According to two former defense attachés to Nigeria, low institutional capacity caused by this tension has an adverse impact on the ability of Nigerian forces to absorb and sustain ACOTA training.

Rwanda's high military capability, on the other hand, may be due in part to Rwanda's tragic recent history of genocide. According to State Department trainers and OSD desk officers, the 1994 genocide has made today's Rwandan population sympathetic to the need for a strong military to maintain internal order, giving the military a pres-

---

[35] These figures are at Purchasing Power Parity, based on Economist Intelligence Unit data.

[36] Patrick J. McGowan, "African Military Coups d'Etat, 1956–2001: Frequency, Trends and Distribution," *Journal of Modern African Studies* (2003) 41: 339–370.

tigious role in society. Furthermore, the current President of Rwanda, Paul Kagame, has a military background as founder of the main Tutsi armed rebel group, and is a strong supporter of the national military.

Lastly, Senegal's relatively stable and democratic history has made it a recipient of extensive aid from donor organizations. DoD officials speculated that this favored aid recipient status may have diminished the Senegalese government's resolve to become self-sufficient, which could account for Senegal's comparatively slow progress toward developing its own cadre of competent peacekeeping trainers.

Overall Assessment

Develop/sustain armed forces. It is not possible to firmly conclude that ACOTA has a positive impact on troop effectiveness in deployment outcomes, due to the lack of data that would enable an analyst to reliably assess the program's causal impact. In particular, there is a lack of systematic data collection that would enable analysts to track troops' pre-training skill-level to their performance in peacekeeping operations. Only in 2007 did the State Department begin to capture individual troop-level data and begin efforts to track troops' careers in order to account for attrition from partner militaries, which some officials estimated to be quite high due in part to HIV/AIDS prevalence in some African countries.

In addition to collecting data at the individual level, it would be helpful for the United States to systematically collect information on deployment outcomes, and to use the resulting information to inform ACOTA training.

U.S. officials have generally been pleased with the performance of Nigerian battalions in Darfur. However, DoD officials have raised concerns about Nigerian battalion skills in Darfur.

Participation in regional stability operations arrangements. The ACOTA program appears to positively influence the battalions that receive ACOTA training, and helps African nations meet the high demand for peacekeepers on their continent, which has increased significantly since the U.N. assumed control of the mission in Darfur and expanded the peacekeeping force there.

Partner-Selection Models

This appendix provides a technical description of the two partner-selection models—coalition/regional and indigenous—used in the exploratory analysis discussed in Chapter Five. Each model is described in terms of its overall structure, as well as its individual attributes, indicators, and scoring methodology.

Coalition/Regional Model

The structure of the coalition/regional model is shown in Figure E.1. The overall score for each potential partner country is derived from three attribute scores that measure the country's capability, willingness, and appropriateness. The overall score and the scores for the three attributes are between zero and one ([0, 1]). In the base case, the attributes are weighted equally (1/3 each) and then summed to give the overall score.

Similarly, the attribute scores are derived from various indicator scores that are also [0, 1]. Again, in the base case, indictors are weighted equally to determine the attribute score.

Indicator scores are derived from data from a variety of authoritative sources that are cited in the reference sheet of the model.

Mapping functions are used to force the indicators to be in the range [0, 1]. These mapping functions convert the extremes of the indicator to zero or one and the middle values to the range [0, 1]. The points that are mapped to zero and one, and the direction of the mapping can be changed in order to conduct exploratory analysis.

Figure E.1
Schematic of the Coalition and Regional Model

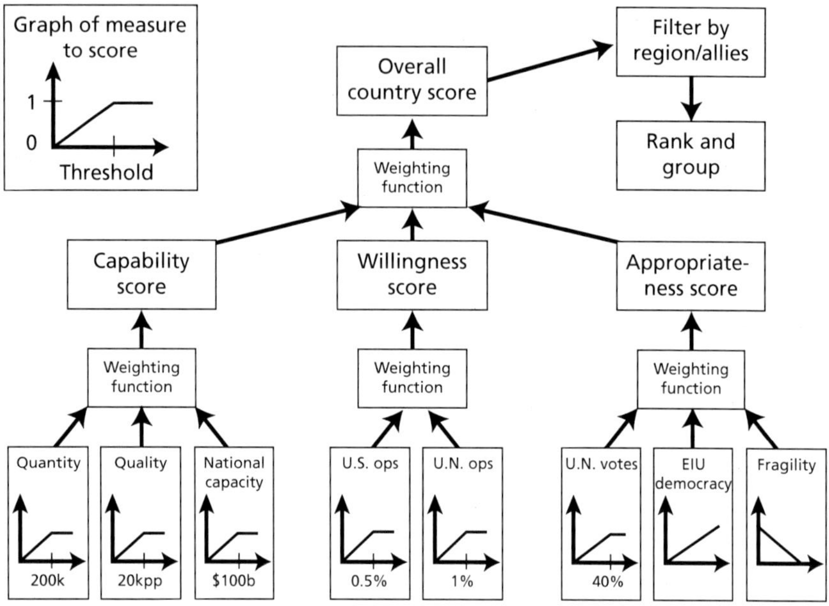

RAND MG942-E.1

To summarize, each set of measurement data is thus converted to an indicator score. The weighted sum of indicator scores results in the attribute score. The weighted sum of the attribute scores provides the overall score.

The overall scores for each country are then tabulated, along with their associated attribute scores. These results can be sorted alphabetically by country or by descending overall score (the higher scores indicate better partners).

For a regional perspective, country results can be filtered by COCOM (a fully operational AFRICOM is assumed) by clicking a button in the spreadsheet interface.

Capability

The capability attribute is derived from three indicators: troop quantity, troop quality, and national capacity in the form of gross domestic product.

Troop quantity consists of the number of personnel in a country's armed forces, excluding paramilitary and reserve forces. Data for this indicator were mostly obtained from the International Institute for Strategic Studies, although *Jane's World Armies* and the U.S. State Department web site were consulted for missing data points.

The number of troops is a measure of a country's ability to contribute to coalition operations. The assumption is that the more troops a country has within its defense establishment, the more troops it can deploy outside its borders. While this is generally true, countries that are already engaged in security activities may be unable to participate in additional operations. An example of such a country is Sri Lanka, which receives a high score in this measure.

The troop quantity score is determined by linearly mapping [0, 200,000] troops to the score [0, 1]. A country with 200,000 or more troops gets a score of one. This force level was deemed a sufficient partner contribution for any stability operations-related coalition mission. A few countries have higher troop counts. However, setting the threshold at the maximum possible level (China's 2.25 million troops) would have artificially deflated the capability scores of other potential partners.

Troop quality is extremely difficult to measure in a comprehensive and consistent way. We developed a proxy indicator: military spending per troop, using information gathered primarily from the International Institute for Strategic Studies and, secondarily, from *Jane's* and Global Security.org. This seemed to be an appropriate surrogate for quality in that many countries devote a large percentage of defense budgets to training troops and paying their salaries. However, we acknowledge that not all military training and personnel are relevant to stability operations. Furthermore, a few countries spend a disproportionate share of their defense budgets on capital acquisition for systems that may not aid stability operations. Given the relatively low training requirements for stability operations, the high threshold for troop quality was set at $20,000 per soldier.

Gross domestic product was included as an indicator of a country's capability to sustain the military and nonmilitary aspects of an external stability operation. For the most part, we relied on World

Bank GDP figures from 2005, the most recent dataset that was nearly complete. The CIA's *World Fact Book* provided additional data. The high threshold for the GDP indicator was set at $100 billion, reflecting our view that a large number of countries probably have the economic capacity to support stability operations.[1]

Willingness

The willingness attribute is composed of the average contributions to recent U.S.-led operations and U.N.-led operations as a percentage of a country's overall force size. Data on country support for U.S.-led operations were taken from a recently published RAND Arroyo report, *Building Partner Capabilities for Coalition Operations.*[2] U.N. deployments were drawn from the United Nations Peacekeeping web site. The latter indicator captures a broad willingness to engage in stability operations, which may be especially meaningful for regional partners.

The fact that willingness, in the case of both indicators, is defined as a percentage of the overall force deployed gives an advantage to small nations. It is easier for a small nation, like Fiji, to deploy a large percentage of its armed forces than it is for larger countries in Western Europe or South Asia. An alternative would to be to simply focus on the total number of troops deployed in out-of-country operations. However, this would favor large countries and confound the willingness variable with the capability variable.

The aforementioned Arroyo report examines foreign partner contributions to eight U.S.-led coalition operations: Operation Iraqi Freedom, Operation Enduring Freedom, International Security Assistance Force, Bosnia, Kosovo, Haiti, Sinai, and Somalia. Although the report includes contributions that did not involve military forces, such as providing overflight rights or temporary bases, this type of assistance did not figure in our willingness calculations since they are not quantifiable in terms of troop numbers. Whenever possible, we replaced entries such as "major fleet unit" and "brigade" with troop estimates.

---

[1] We did not attempt to convert our GDP data to Purchasing Power Parity terms.

[2] Jennifer D.P. Moroney et al., MG-635, 2007, pp. 89–95.

Our source of data for foreign participation in U.S.-led coalition operations is not entirely suitable as the basis for a willingness measure. First, the study only covers recent operations, some of which do not have a large stability operations component. Second, these operations are not distributed evenly around the globe. A country is more likely to be a willing stability operations participant if its national interests are at stake, and for most countries these interests are highly correlated with the proximity of the conflict to their own territory. The under-representation of U.S.-led operations in Asia and South America in this dataset may introduce a bias into our willingness measure against countries in these regions. However, such a bias may be justified if we assume that future operations will take place in the same regions where they have occurred in the recent past.

Thresholds for deployment indicators were initially set to provide a reasonable spread of willingness scores and to avoid a bias in favor of countries with large armed forces. The high threshold was 0.5 percent of total military forces for contributions to U.S.-led coalition operations and 1 percent of the total force for contributions to U.N.-led peacekeeping operations.

U.N. coalition information was collected from the U.N. Department of Peacekeeping Operations (DPKO) web site. This source includes only U.N. operations directed and supported by the DPKO. The data were sampled in six-monthly intervals between October 2003 and April 2007. We used the total numbers deployed to calculate the average number of troops deployed at any time. We then converted this figure to a percentage of troops deployed, borrowing the total troop numbers from the capability attribute.

Like the U.S. data, the U.N. deployment information covers a relatively short time period and may be unduly influenced by current events. For example, during the DPKO sample period, the United Kingdom was heavily involved in Iraq and Afghanistan which, arguably, may have prevented it from deploying to U.N. operations. Since the United Kingdom exceeds the maximum threshold for U.S.-led operations by a factor of four, our default scoring system penalizes the United Kingdom in terms of willingness by weighing contributions to U.N. and U.S.-led operations equally.

Additionally, some multinational peacekeeping operations do not fall under the mandate of the DPKO. For example, the Australian-led operation in East Timor is not included in our dataset, even though it is sanctioned by the U.N.

Neither willingness indicator takes into consideration the motives of the countries participating in U.S.- and/or U.N.-led operations. Although some countries may provide troops out of a sense of duty or loyalty, other countries may be primarily motivated by economic incentives. This would seem to be the case for several relatively poor countries such as Fiji, which has the highest score for willingness of any nation in our sample.

Even if data were available to distinguish among various motives, we did not believe this information should be a major element in the willingness measure. There may be cases in which the United States would be disinclined to fund the participation of a country in a coalition operation despite the latter's apparent readiness to deploy troops abroad. However, the rationale for such a decision would likely be captured in the appropriateness or capability attributes of our coalition/regional model.

Some countries receive very low willingness scores despite being allies of the United States. With a willingness score of zero, Israel is the prime example. Our model cannot explain whether this result reflects a true lack of willingness on Israel's part to participate in coalition operations or whether the United States and the U.N. have refrained from requesting Israeli participation for political reasons. This is an example where our quantitative, macro-level analysis should be supplemented by qualitative, country-specific information.

Appropriateness

Appropriateness was included as a litmus test for potential coalition partners because certain capable and willing countries may not meet the grade for political and/or strategic reasons. Furthermore, some less-capable or previously unwilling countries may make acceptable BPC candidates given their political and/or strategic alignment with the United States. Appropriateness is a composite measure of democratization, U.N. voting, and fragility. Respectively, they indicate whether a country is politically and/or ideologically similar to the United States, has a similar international outlook as does the United States, and is domestically stable.

The democratization indicator is derived from an Economist Intelligence Unit index, which has five subcomponents: electoral process and pluralism; civil liberties; government functioning; political participation; and political culture. Composite scores based on these subcomponents are used to rank countries and group them into the following categories: full democracies, flawed democracies, hybrid regimes, and authoritarian regimes. The index provides a snapshot of the state of democracy in 165 countries and two territories.

The strength of the democratization indicator is that it is based on a broader concept of democracy than simply holding fair elections; it considers the social and cultural underpinnings of democratic development. However, the main disadvantage of this indicator is that the Economist Intelligence Unit index results differ from those produced by other democratization indices. In most cases, these differences are slight, but they can be significant. Although disconcerting, this disparity is not unexpected given the difficulty of objectively measuring such a complex social phenomenon.

The U.N. voting indicator attempts to capture the degree of strategic affinity that exists between foreign countries and the United States. As tabulated by the U.S. State Department, this measure includes individual country votes in the U.N. General Assembly and excludes consensus votes. We chose to rely on a country's overall voting behavior rather than its record on DOS-designated "important votes." The latter category seemed too narrowly constructed for our purposes, focusing on a small number of politically charged issues, many involving Israel and the Palestinians.

U.N. votes are not a perfect measure of policy agreement with the United States. This is particularly evident as the overall level of support for the U.S. voting position decreases. Even some Western European countries score surprisingly low in terms of the percentage of their U.N. votes that align with those of the United States. As a result, the high threshold for this measure was set at 40 percent.

Fragility was measured using the Failed States Index compiled by the Fund for Peace. Using the Conflict Assessment System Tool, this index ranks 177 states according to twelve social, political and economic indicators. Ratings reflect a state's vulnerability to collapse or conflict, to include: loss of control over its territory, loss of its monopoly on the legitimate use of force, an erosion of its authority to make collective decisions, an inability to provide essential public services, and an inability to interact with other states as a full member of the international community.

Regional Analysis

To make it easier to examine the rankings of potential partners by region, each country in the coalition/regional model was tagged with the COCOM to which it belonged. For analytical purposes, we assumed a fully operational AFRICOM.

This DoD-oriented regional categorization system has pluses and minuses. A division of countries along ethnic or religious lines might have provided combinations of countries with a greater number of shared interests. However, it is not possible to decisively draw such lines, nor would the large number of possible divisions prove helpful. COCOM areas of responsibility have the advantage of being well-defined and of direct relevance when it comes to implementing partner-selection recommendations. In addition, the COCOMs are reasonably well-aligned with many other regional frameworks.

All categorization schemes that place each country in only one region suffer from similar seam issues. This was addressed in our study by exploring the rankings of certain major countries (those that fall in one COCOM area of responsibility but are of high importance to others) in more than one regional context. These countries are shown in Table E.1.

Table E.1
Countries in the Seam Between COCOMs

Country	Actual COCOM	Explored COCOM
Turkey	EUCOM	CENTCOM
Egypt	CENTCOM	AFRICOM
Mexico	NORTHCOM	SOUTHCOM
India	PACOM	CENTCOM

Distribution of Country Scores

The purpose of the coalition/regional model was to provide a quick and broad assessment of countries' suitability to be stability operations partners.

As Figure E.2 shows, the distribution of country scores is not uniform. The majority of countries are low scoring (121 score less than 0.5). These countries form a bell-shaped curve centered between 0.2 and 0.3. Of the remaining 40 nations, 26 are in the range 0.6–0.8.

Indigenous Model

With only two major attributes and three indicators, the indigenous model is similar to, but simpler than, the coalition/regional model (see Figure E.3). The fragility attribute has just one indicator, which is

Figure E.2
Distribution of Country Scores

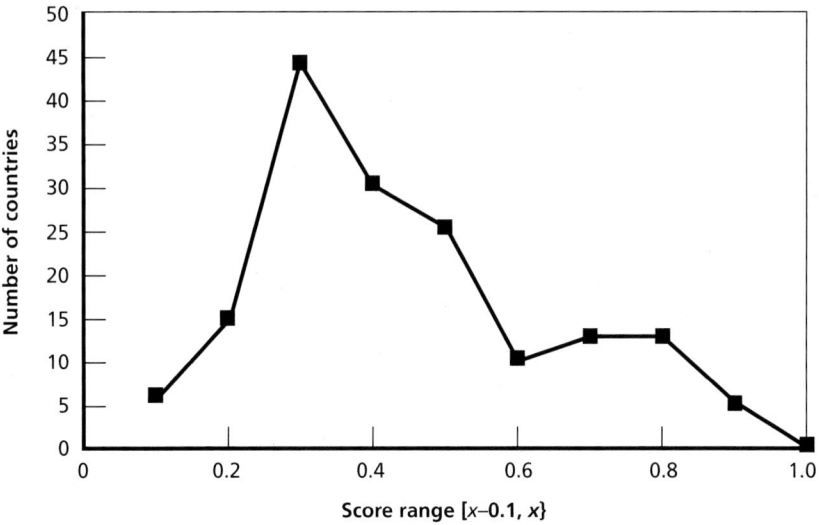

Figure E.3
Indigenous Model

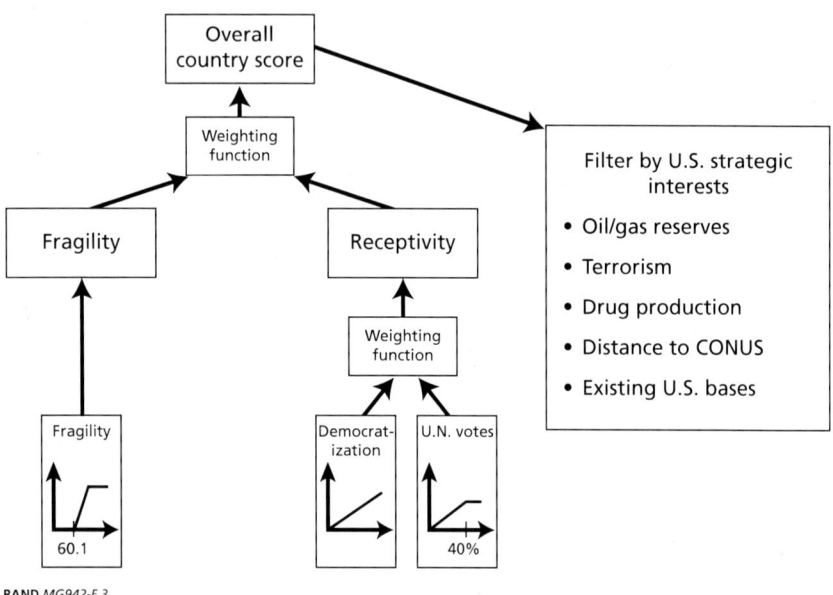

RAND *MG942-E.3*

derived from the Failed States Index. The receptivity attribute has two indicators: U.N. votes and the Economist Intelligence Unit's Democratization index. These measures are also used in the coalition/regional model, although the rationale for selecting them is different.

The default weightings used to determine the overall score in the indigenous model are 2/3 (fragility) and 1/3 (receptivity). This varies from the uniform weights used elsewhere in both models. This reflects the idea that the purpose of this model is to find fragile countries that are also receptive. Of course, analysts have the option to change this weighting function.

Structurally, the indigenous model differs from the coalition model in its use of five U.S. strategic interests as post-processing filters, which come into play after countries have been ranked according to their weighted attributes. These include

- Maintaining access to energy sources.
- Countering international terrorism.
- Countering illicit drug production.
- Protecting the near abroad.
- Protecting overseas bases.

Using the appropriate buttons on the summary spreadsheet, the model allows the analyst to sort countries by particular strategic interest or to find countries in which the United States has any strategic interest.

Fragility

As in the regional/coalition model, the indigenous model's fragility attribute is also based on the Fund for Peace's Failed States Index, which identifies four levels of fragility: alert, warning, moderate, and sustainable.

However, the indigenous model diverges from the Failed States Index in one way. In the Failed States Index, the moderate level of fragility is associated with countries on par with the United States, which we do not consider realistic candidates for indigenous BPC support. Therefore, we have conflated the "moderate" and "sustainable" categories into a single category.

Democratization

Democratization (as defined by the Economist Intelligence Unit) was selected as an indicator of receptivity based on the assumption that democratic governments are more willing and able to effect reform. We acknowledge that this may not always be the case. In some instances, an authoritarian government, with a strong interest in maintaining good relations with the United States, may be more capable of executing significant political changes than a fractious democratic regime, especially if these changes do not meet with the approval of important domestic constituencies. However, democratic governments have a better record of maintaining a course of reform over the long term without provoking widespread social disruption.

U.N. Votes

The U.N. votes indicator reflects countries' receptivity to U.S. foreign policy goals. Our assumption is that the more willing a country is to vote with the United States in the U.N., the more likely it is to accept U.S. aid and use it in a way that conforms to U.S. preferences. Although this assumption may not be valid at all times for every country, a country's U.N. voting record tends to highlight countries with regimes that are actively hostile to the United States and thus would not be trustworthy recipients of American assistance. As in the coalition model, the high threshold for this indicator in the indigenous model was set at 40 percent, which roughly corresponds to the level of voting alignment between the United States and its traditional European allies.

U.S. Strategic Interests

Because of the large number of countries in need of security assistance, we considered it sensible to develop a mechanism for segregating those in which the United States has compelling strategic interests. Although we do not claim that our model fully incorporates these interests (some of which are politically contentious and/or difficult to quantify), we believe that the five we have chosen approximate the range of interests that are important to most U.S. policymakers.

Given the uncertain fidelity of our data, countries were assessed as either being of strategic interest to the United States in a particular area (in which case they received a score of 1) or not (in which case they received a score of 0). No intermediate scores were calculated. Additionally, strategic interests were not combined in any way. Our summary of U.S. strategic interests indicates only whether or not a country met at least one of the criteria.

The following sections provide more information on the five strategic interests that we selected for our indigenous model.

Oil/Natural Gas. Energy security is a key concern for the United States and an important component of the U.S. interest in the Middle East, Africa, and Central Asia. To assess the extent of this particular interest, we used a list of countries' proven oil and natural gas reserves compiled by the U.S. Department of Energy. This list contains three estimates for each country. So as not to unduly restrict our grouping of strategic countries, we used the largest of these estimates except when there was one estimate that deviated wildly from the other two. We then calculated the average price over a two-year period for oil and gas, and combined the results in dollar terms.

After examining various options, we chose to set the minimum oil/gas threshold for a strategic country at one percent of the total production value of the countries in the indigenous model. According to this definition, the United States has a strategic energy interest in the countries shown in Table E.2.

Terrorism. The United States has a general interest in reducing the incidence of terrorism around the world. That said, the country's most pressing interest is in preventing terrorism directed at U.S. citizens. For this reason, we chose to use as our terrorism indicator countries in which terrorist attacks have been directed against U.S. civilians and/or U.S. property. An alternative indicator might have been the national origin of the perpetrators of terrorist strikes against Americans. However, this information was not available in the large majority of cases.

The National Counterterrorism Center database includes 123 incidents in 43 countries that meet our U.S. targeting criteria. The incidents that were cataloged included the following types of terrorist attacks: Islamic extremism, political violence, and environmental terrorism. Establishing a minimum threshold of two attacks reduces

Table E.2
Countries with Significant Energy Reserves That Are of Strategic Interest to the United States

U.S. Strategic Interests: Energy Reserves	
Iraq	Saudi Arabia
Nigeria	United States
Russian Federation	Libya
Venezuela	Kuwait
China	Canada
Iran	Qatar
Kazakhstan	Norway
Algeria	United Arab Emirates

the number of countries in which the United States has a strategic interest due to terrorism to the 20 countries shown in Table E.3.

There are some countries that have been associated with terrorism that are missing from the above list. Iran, Syria, and North Korea are prominent examples. Although these countries have been accused of sponsoring the activities of terrorist groups in other countries, the National Counterterrorism Center database does not show any terrorist attacks directed against U.S. targets that occurred on their soil. In any event, these alleged state sponsors of terrorism are unlikely recipients of U.S. BPC assistance.

Illegal Drugs. The three elements of the drug problem that we could have addressed were production, trafficking, and usage. We concluded that usage was a domestic issue and therefore not relevant to our model. In addition, we considered our proximity measure covered the countries most implicated in U.S.-oriented drug trafficking. This left drug production as the basis for our strategic interest indicator. Although comprehensive data on drug production per country is lacking, the vast majority of the world's coca and opium (the most important raw ingredients of illegal drugs) comes from a few well-known countries. The following countries were identified in the U.N. *2007 World Drug Report*:[3]

- Afghanistan
- Bolivia
- Colombia
- Laos
- Mexico
- Myanmar
- Pakistan
- Peru.

Geographic Proximity. Despite its status as a global power, the United States' interest in certain countries is very much influenced by their proximity to the U.S. homeland. A classic example would be

---

[3] United Nations Office on Drugs and Crime, *2007 World Drug Report*.

Table E.3
Countries with Terrorist Attacks Against U.S. Targets

Terror Attack Locations	
Iraq	Philippines
Afghanistan	Serbia
Israel (Includes	Egypt[a]
Occupied Territories)[a]	Thailand
Pakistan[a]	India
Bangladesh	Jordan
Nigeria	Greece
Nepal	Saudi Arabia[a]
Turkey	Italy
Indonesia[a]	Argentina

[a] Countries with two or more Islamic extremist attacks.

Cuba, 90 miles off the Florida coast, whose government allowed the deployment of Soviet nuclear weapons during the early 1960s, provoking the most serious crisis of the Cold War. More recently, geography has enabled other problems to spread to American shores. These problems include the trafficking of illegal drugs mentioned above, as well as uncontrolled migration as a result of economic privation, civil war, political repression, and natural disasters in neighboring countries.

For the purpose of our indigenous analysis, countries located in North America, Central America, and the Caribbean were deemed to be strategically close to the United States, as were South American countries bordering the Caribbean, in particular, Colombia and Venezuela.

U.S. Bases. The final U.S. strategic interest that we examined was countries hosting U.S. military bases. The list of such countries was extracted from the FY 2006 report issued by the Office of the Deputy Under Secretary of Defense (Installations and Environment). For our analysis, we included any country with an active U.S. military installation (i.e., provider of employment), with the exclusion of Iraq and Afghanistan, whose operational bases were considered temporary. Selected countries with U.S. bases are shown in Table E.4.

Table E.4
Location of United States Overseas Bases

Countries With U.S. Bases		
Australia	Germany	Oman
Bahrain	Greece	Peru
Belgium	Indonesia	Portugal
Canada	Italy	Qatar
Colombia	Japan	Singapore
Cuba	Kenya	Spain
Denmark	Korea, Republic of	Turkey
Ecuador	Kuwait	United Arab Emirates
Egypt	Netherlands	United Kingdom
France	Norway	

Bibliography

Baldauf, Scott, "U.S. Steps Up Its Military Presence in Africa," *Christian Science Monitor*, October 2, 2007.

Bates, Robert H., et al., *Political Instability Task Force Report, Phase IV Findings*, November 18, 2003.

Casey, General George W., *Opening Remarks, Senate Armed Service Committee*, 26 February 2008. As of October 10, 2008:
http://www.army.mil/-speeches/2008/02/26/7823-opening-remarks-senate-armed-services-committee/

Defense Security Cooperation Agency, *Security Assistance Management Manual*, DoD 5105.38-M, 2007. As of January 2010:
http://www.dsca.mil/SAMM/

Defense Security Cooperation Agency, FAQ, 2007. As of January 2010:
http://www.dsca.mil/PressReleases/faq.htm

Department of Defense, Air Land Sea Application Center, *Multi-Service Tactics, Techniques, and Procedures for Conducting Peace Operations*, FM 3-07.31, October 2003.

Department of Defense, *Quadrennial Defense Review Building Partnership Capacity Execution Roadmap*, Washington, D.C.: Office of the Secretary of Defense and the Joint Staff J-5, May 2006.

Department of Defense Directive 3000.05, *Military Support for Stability, Security, Transition, and Reconstruction Operations*, November 28, 2005. As of January 2010:
http://usacac.army.mil/cac2/sfa/Repository/DoD_Dir-FSFA.pdf

Department of Defense Instruction 3000.05, *Stability Operations*, September 16, 2009. As of January 2010:
http://www.dtic.mil/whs/directives/corres/pdf/300005p.pdf

Department of Defense, Joint Forces Command, *Military Support to Stabilization, Security, Transition, and Reconstruction Operations Joint Operating Concept*, December 2006.

Department of Energy, Energy Information Administration, "World Proved Reserves of Oil and Natural Gas, Most Recent Estimates," spreadsheet. As of January 2010:
http://www.eia.doe.gov/emeu/international/reserves.xls

Department of Energy, Energy Information Administration, "World Crude Oil Prices." As of January 2010:
http://tonto.eia.doe.gov/dnav/pet/pet_pri_wco_k_w.htm

Department of Energy, Energy Information Administration, "Oil: Crude and Petroleum Products Explained." As of January 2010:
http://tonto.eia.doe.gov/energyexplained/index.cfm?page=oil_home#tab2

Department of State, *Foreign Assistance Standardized Program Structure and Definitions*, 20 October 2006. Updated 15 January 2010. As of February 2010:
http://www.state.gov/documents/organization/136594.pdf

Department of State, Bureau of International Organization Affairs, report to Congress, *Voting Participation in the U.N. 2006*, "Part IV: General Assembly: Important Votes and Consensus Actions." As of January 2010:
http://www.state.gov/documents/organization/82643.pdf

Department of State, Bureau of Political-Military Affairs, web site. As of January 2010:
http://www.state.gov/t/pm/index.htm

Department of State, Director of U.S. Foreign Assistance, "Standardized Program Structure and Definitions." As of February 2010:
http://www.state.gov/f/c24132.htm

Department of State, Office of the Coordinator for Reconstruction and Stabilization, web site. As of January 2010:
http://www.state.gov/s/crs/

Department of State, Office of the Coordinator for Reconstruction and Stabilization, *Reconstruction and Stabilization Essential Tasks,* April 2005. As of January 2010:
http://www.crs.state.gov/index.cfm?fuseaction=public.
display&id=10234c2e-a5fc-4333-bd82-037d1d42b725
PDF of *Post-Conflict Reconstruction Essential Tasks* available at:
http://www.crs.state.gov/index.cfm?fuseaction=public.display&shortcut=J7R3

Economist Intelligence Unit, web site. As of January 2010:
http://www.eiu.com

France Diplomatie, "France in the U.N. System," February 21, 2006. As of January 2010:
http://www.diplomatie.gouv.fr/en/france-priorities_1/international-organizations_1100/france-in-the-un-system_3281/index.html.

Fund for Peace, "Failed States Index." As of January 2010:
http://www.fundforpeace.org/web/index.php?option=com_content&task=view&id
=229&Itemid=366

Gates, Robert M., "A Balanced Strategy: Reprogramming the Pentagon for a New
Age," *Foreign Affairs*, January/February 2009. As of March 23, 2009:
http://www.foreignaffairs.com/articles/63717/robert-m-gates/a-balanced-strategy

George D. Marshall European Center for Strategic Studies, mission statement, 15
September 2009. As of January 2010:
http://www.marshallcenter.org/mcpublicweb/en/nav-mc-about-mission.html

German Federal Ministry of Defense web site. As of January 2010:
http://www.bmvg.de/portal/a/bmvg

GlobalSecurity.org web site. As of January 2010:
http://www.globalsecurity.org/

Headquarters, Department of the Army, *Civil Affairs Operations*, FM 41-10,
February 2000.

Headquarters, Department of the Army, *Mine/Countermine Operations*, FM 20-32,
October 2002.

Headquarters, Department of the Army, *Army Universal Task List*, FM 7-15, July
2006.

Headquarters, Department of the Army, *Counterinsurgency*, FM 3-24, December
2006.

Headquarters, Department of the Army, *Operations*, Field Manual 3-0, February
2008.

Headquarters, Department of the Army, *Stability Operations*, Field Manual 3-07,
October 2008. (Earlier edition was dated February 2003.) As of January 2010:
http://usacac.army.mil/cac2/Repository/FM307/FM3-07.pdf

Helmonoed-Romer Heitman, "SADC Launches Standby Brigade," *Jane's Defence
Weekly*, Vol. 44, No. 35, August 29, 2007, p. 17.

International Force East Timor, informational web site. As of January 2010:
http://pandora.nla.gov.au/parchive/2000/S2000-Nov-7/easttimor.defence.gov.au/
index.html.

Jane's Information Group, "South Africa: External Affairs," Jane's Sentinel
Security Assessment—Southern Africa, March 2006.

Jane's World Armies. As of January 2010:
http://www2.janes.com/K2/blvl2.jsp?Category=Countries&SelPub=jwar

Joint Chiefs of Staff, *Joint Operations*, JP 3-0, February 2008.

Joint Chiefs of Staff, *Peace Operations*, JP 3-07.3, October 17, 2007.

Joint Chiefs of Staff, *Universal Joint Task List,* Chairman of the Joint Chiefs of Staff Manual 3500.04D, August 2005.

Kofi Annan International Peacekeeping Training Centre, web site. As of January 2010:
http://www.kaiptc.org/home

Marquis, Jefferson P., Richard E. Darilek, Jasen J. Castillo, et al., *Assessing the Value of U.S. Army International Activities,* Santa Monica, CA: RAND Corporation, MG-329-A, 2006.
http://www.rand.org/pubs/monographs/MG329/

Marshall, Monty G., and Ted Robert Gurr, *Peace and Conflict 2005: A Global Survey of Armed Conflicts, Self-Determination Movements, and Democracy,* College Park, MD: Center for International Development and Conflict Management, 2005. As of January 2010:
http://www.cidcm.umd.edu/publications/papers/peace_and_conflict_2005.pdf

Marshall, Monty G., and Keith Jaggers, "Polity IV Project: Political Regime Characteristics and Transitions, 1800–2008." As of February 2010:
http://www.systemicpeace.org/polity/polity4.htm

McDoom, Opheera, "Ethiopia Pledges 5,000 Peacekeepers to Darfur," *Reuters,* October 4, 2007.

McGowan, Patrick J., "African Military Coups d'État, 1956–2001: Frequency Trends and Distribution," *Journal of Modern African Studies,* Vol. 41, Issue 3, 2003, pp. 339–370.

Moroney, Jennifer D.P., et al., *Building Partner Capabilities for Coalition Operations,* Santa Monica, CA: RAND Corporation, MG-635-A, 2007.
http://www.rand.org/pubs/monographs/MG635/

Moroney, Jennifer D.P., Adam Grissom, and Jefferson P. Marquis, *A Capabilities-Based Strategy for Army Security Cooperation,* Santa Monica, CA: RAND Corporation, MG-563-A, 2007.
http://www.rand.org/pubs/monographs/MG563/

National Counterterrorism Center, "Worldwide Incidents Tracking System," database of terrorist incidents. As of January 2010:
http://wits.nctc.gov/

National Defence and the Canadian Forces, "Directorate Military Training and Cooperation: Background." As of February 2010:
http://www.forces.gc.ca/admpol/newsite/mtcpbackground-eng.html

National Guard, International Affairs Division web site. As of February 2010:
http://www.ng.mil/jointstaff/j5/ia/default.aspx

Pincus, Walter, "Taking Defense's Hand Out of State's Pocket," *Washington Post,* July 9, 2007, p. A17.

Ploch, Lauren, *Africa Command: U.S. Strategic Interests and the Role of the U.S. Military in Africa,* Congressional Research Service Report RL34003, October 2, 2009. As of January 2010:
http://www.fas.org/sgp/crs/natsec/RL34003.pdf

Regional Assistance Mission to Solomon Islands, web site. As of January 2010:
http://www.ramsi.org

Serafino, Nina, "The Global Peace Operations Initiative: Background and Issues for Congress," Congressional Research Service Report RL32772, updated March 19, 2009. As of January 2010:
http://www.fas.org/sgp/crs/misc/RL32773.pdf

South African Development Community, web site. As of January 2010:
http://www.sadc.int/

Turkey's Centre of Excellence, Defence Against Terrorism, web site. As of January 2010:
http://www.tmmm.tsk.tr/

Tyson, Ann Scott, "U.S. to Raise 'Irregular War' Capabilities," *Washington Post,* December 4, 2008.

United Kingdom Department for International Development, *The Global Conflict Prevention Pool: A Joint UK Government Approach to Reducing Conflict.* As of February 2010:
http://www.dfid.gov.uk/Documents/publications/global-conflict-prevention-pool.pdf

United Kingdom Ministry of Defence, "The British Ministry Advisory and Training Team (Czech Republic)." As of February 2010:
http://www.mod.uk/DefenceInternet/AboutDefence/WhatWeDo/DoctrineOperationsandDiplomacy/BMATTCZ/

United Kingdom Ministry of Defence, Fact Sheet, "British Military Advisory Training Team West Africa." As of January 2010:
http://www.operations.mod.uk/africa/bmattwa.htm

United Kingdom Ministry of Defence, Fact Sheet, "International Military Assistance Training Team Sierra Leone." As of January 2010:
http://www.operations.mod.uk/africa/imattsl.htm

United Nations, "Monthly Summary of Contributors of Military and Police Personnel." As of January 2010:
http://www.un.org/Depts/dpko/dpko/contributors

United Nations Office on Drugs and Crime, *2007 World Drug Report.* As of January 2010:
http://www.unodc.org/india/world_drug_report_2007.html

United Nations RECAMP Programme, "Field Peacekeeping Training." As of ~uary 2010:

http://www.un.int/france/frame_anglais/france_and_un/france_and_
peacekeeping/field_pktraining_eng.htm

United States Africa Command web site. As of January 2010:
http://www.africom.mil/

USAID, "Fragile States Strategy," Washington, D.C.: USAID, PD-ACA-999,
January 2005.

"U.S. Defense Chief Urges Greater Use of 'Soft Power,'" Agence France-Presse,
November 26, 2007. As of November 20, 2008:
http://afp.google.com/article/ALeqM5i1-FVKEwqEMcOu7agak_FXdCnQtw

White House, National Security Presidential Directive 44, *Management of
Interagency Efforts Concerning Reconstruction and Stabilization,* December 7, 2005.
As of January 2010:
http://www.fas.org/irp/offdocs/nspd/nspd-44.html

White House, President George W. Bush, *The National Security Strategy,*
September 2002. As of December 2009:
http://georgewbush-whitehouse.archives.gov/nsc/nss/2002/

World Bank, "Data and Statistics: World Development Indicators 2009." As of
February 2010:
http://go.worldbank.org/0ROQCBCZG0